The 2nd International Conference of International Researchers of the Education for Environmental Citizenship 2022

The 2nd International Conference of International Researchers of the Education for Environmental Citizenship 2022

Editors

Andreas Ch. Hadjichambis
Pedro Reis
Marie-Christine Knippels
Audronė Telešienė
Daphne Goldman
Demetra Paraskeva-Hadjichambi
Jan Činčera
Kateřina Jančaříková

MDPI • Basel • Beijing • Wuhan • Barcelona • Belgrade • Manchester • Tokyo • Cluj • Tianjin

Editors

Andreas Ch. Hadjichambis
Cyprus Centre for Environmental Research and Education CYCERE
Cyprus

Pedro Reis
Instituto de Educação da Universidade de Lisboa
Portugal

Marie-Christine Knippels
Utrecht University
The Netherlands

Audronė Telešienė
Kaunas University of Technology
Lithuania

Daphne Goldman
Beit Berl Academic College
Israel

Demetra Paraskeva-Hadjichambi
Cyprus Centre for Environmental Research and Education CYCERE
Cyprus

Jan Činčera
Masaryk University
Czech Republic

Kateřina Jančaříková
Charles University
Czech Republic

Editorial Office
MDPI
St. Alban-Anlage 66
4052 Basel, Switzerland

This is a reprint of articles from the Proceedings published online in the open access journal *Environmental Sciences Proceedings* (ISSN 2673-4931) (available at: https://www.mdpi.com/2673-4931/14/1).

For citation purposes, cite each article independently as indicated on the article page online and as indicated below:

LastName, A.A.; LastName, B.B.; LastName, C.C. Article Title. *Journal Name* **Year**, *Volume Number*, Page Range.

ISBN 978-3-0365-4071-9 (Hbk)
ISBN 978-3-0365-4072-6 (PDF)

Cover image courtesy of Xiaoxiao Wang

Contents

Preface to "The 2nd International Conference of International Researchers of the Education for Environmental Citizenship 2022"

Contemporary environmental challenges require a comprehensive and holistic education that is change-oriented. Such education could contribute to empowering capable environmental citizens who could bring about those changes in the environment and society, protecting the environment and achieving sustainability. This is the direction of Education for Environmental Citizenship, which is the theme of the International Conference iREEC 2022. The iREEC 2022 International Conference is intellectually stimulating, enriching our scientific discussions with a visionary spirit, for a more sustainable future of the global environment and the global society. We also hope that iREEC 2022 will lead to an innovative, active, and effective scientific community of International Researchers focusing on Education for Environmental Citizenship.

It is a great honor for the Cyprus Centre for Environmental Research and Education (CYCERE) to hold such an important International Conference on Education for Environmental Citizenship in Europe in the framework of the COST Action "European Network for Environmental Citizenship". We have selected six topics that will include oral presentations, an interactive poster session, and a symposium to create the perfect conditions for an inspirational scientific environment to discuss Environmental Citizenship.

Acknowledgments

Funded by the Horizon 2020 Framework Programme of the European Union

This publication is based upon work from the COST Action European Network for Environmental Citizenship ENEC CA16229, supported by COST (European Cooperation in Science and Technology).

COST (European Cooperation in Science and Technology) is a funding agency for research and innovation networks. Our Actions help to connect research initiatives across Europe and enable scientists to grow their ideas by sharing them with their peers. This boosts their research, career, and innovation.

www.cost.eu

Andreas Ch. Hadjichambis, Pedro Reis, Marie-Christine Knippels, Audronė Telešienė, Daphne Goldman, Demetra Paraskeva-Hadjichambi, Jan Činčera, and Kateřina Jančaříková

Editors

environmental sciences proceedings

Editorial

Statement of Peer Review †

Andreas Hadjichambis

Cyprus Centre for Environmental Research and Education CYCERE, Limassol 3304, Cyprus; a.hadjichambis@cytanet.com.cy

† Presented at the 2nd International Conference of International Researchers of the Education for Environmental Citizenship 2022, 10–11 March 2022; Available online: https://enec-cost.eu/ireec22/.

In submitting conference proceedings to *Environment Sciences Proceedings*, the volume editors of the proceedings certify to the publisher that all papers published in this volume have been subjected to peer review administered by the volume editors. Reviews were conducted by expert referees to the professional and scientific standards expected of a proceedings journal.

- Type of peer review: double-blind
- Conference submission management system: NA
- Number of submissions sent for review: 42
- Number of submissions accepted: 21
- Acceptance rate (number of submissions accepted/number of submissions received): 50%
- Average number of reviews per paper: 2
- Total number of reviewers involved: 84

Citation: Hadjichambis, A. Statement of Peer Review. *Environ. Sci. Proc.* **2022**, *14*, 21. https://doi.org/10.3390/environsciproc2022014021

Published: 12 April 2022

environmental sciences proceedings

MDPI

Proceeding Paper

The Impact of a General Elective Course on Sustainability of the Environmental Citizenship of Undergraduate Students †

Jelle Boeve-de Pauw [1,2], Audronė Telešienė [3,*], Daphne Goldman [4] and Ralph Hansmann [5]

1 Research Unit Edubron, Department of Training and Education Sciences, Faculty of Social Sciences, University of Antwerp, 2000 Antwerp, Belgium; jelle.boevedepauw@uantwerpen.be or j.n.a.boeve-depauw@uu.nl
2 Freudenthal Institute, Faculty of Science, Utrecht University, 3584 CC Utrecht, The Netherlands
3 Civil Society and Sustainability Research Group, Faculty of Social Sciences, Arts and Humanities, Kaunas University of Technology, 44249 Kaunas, Lithuania
4 Department of Environmental Science and Agriculture, Faculty of Education, Beit Berl College, Kfar Saba 44905, Israel; dafnag@beitberl.ac.il
5 Swiss Federal Institute for Forest, Snow and Landscape Research (WSL), 8903 Birmensdorf, Switzerland; hansmann@env.ethz.ch
* Correspondence: audrone.telesiene@ktu.lt; Tel.: +370-68465096
† Presented at the 2nd International Conference of International Researchers of the Education for Environmental Citizenship 2022, 10–11 March 2022. Available online: https://enec-cost.eu/ireec22/.

Abstract: The aim is to evaluate the effect of a sustainable development-oriented course on the environmental citizenship attributes of undergraduate students at a technical university. We discuss the pedagogical and didactic aspects of the course and its linkages to sustainable development education, define the elements of the educational intervention, and present the case study from Kaunas University of Technology. The presentation discusses results from a randomized pre-group -post-group quasi-experimental survey. The results prove the positive effect of the studied educational intervention on the environmental citizenship of undergraduate students.

Keywords: environmental citizenship; education for environmental citizenship; pedagogical approaches in environmental education; higher education; educational evaluation

Citation: Boeve-de Pauw, J.; Telešienė, A.; Goldman, D.; Hansmann, R. The Impact of a General Elective Course on Sustainability of the Environmental Citizenship of Undergraduate Students. *Environ. Sci. Proc.* **2022**, *14*, 19. https://doi.org/10.3390/environsciproc2022014019

Academic Editors: Andreas Ch. Hadjichambis, Pedro Reis, Marie-Christine Knippels, Demetra Paraskeva-Hadjichambi, Jan Cincera and Kateřina Jančaříková

Published: 4 April 2022

Publisher's Note: MDPI stays neutral with regard to jurisdictional claims in published maps and institutional affiliations.

1. Rationale

University students are a crucial audience for environmental citizenship education [1] due to their anticipated roles within citizenry and work environments. Taking a primary interest in active environmental citizenship, we aim to evaluate an educational intervention designed to foster environmental citizenship among undergraduate students at a technological university in Lithuania. We explore the impact of a unique undergraduate sustainable development course on students by implementing a recently developed instrument for measuring the dimensions of environmental citizenship [2]. The instrument is comprised of items measuring past, present, and future actions as environmental citizens and agents of change, as well as the knowledge, conceptions, skills, attitudes, and values for environmental citizenship.

2. Research Question

The group of researchers implemented a quasi-experimental survey design to answer questions about the reliability and validity of the environmental citizenship questionnaire, about the environmental citizenship characteristics of first-year undergraduate students in a large technical Lithuanian university, as well as the impact of the designed educational intervention. These questions are answered, and the survey results are discussed in detail in a separate article [3]. This abstract from the iREEC 2022 conference presentation will further discuss the following research question:

What is the impact of students' participation in a general elective course 'Sustainable Development' on their environmental citizenship?

The research question was answered while testing a general overall hypothesis: the general elective course focusing explicitly on sustainable development (by implementing it via learner-oriented pedagogical methods identified as effective in developing sustainable development competences) will foster increased environmental citizenship in participating students.

3. Methodology

Using a randomized pre-group/post-group quasi-experimental survey design, we explore the students' environmental citizenship before and after the intervention course 'Sustainable Development', compared to students in a general elective course 'Media Philosophy'. Students were divided into two main groups (experimental vs. control) and then randomly assigned to either a pre- or post-intervention survey, resulting in four groups with unique respondent membership. Each of the nine subscales of the environmental citizenship questionnaire showed excellent reliability (alpha values ranging between 0.830 and 0.930); the entire instrument also had excellent internal consistency (0.929).

4. Findings

ANOVA showed that, when comparing the four groups for each of the EC scales, no main effects were present (all $p > 0.05$), suggesting no effect of the intervention on the subscales. However, when the four groups were compared based on the overall EC construct (overall mean of all items of all scales), the results indicate meaningful and significant differences ($p = 0.016$). Two-by-two post hoc analyses showed, as expected, no differences between the pre-intervention group and the pre-control group, and between the pre-control group and the post control group (all $p > 0.05$). However, as hypothesized, the post-intervention group displayed significantly higher EC than the pre-intervention group ($p = 0.007$), and significantly higher EC than the post-control group ($p = 0.025$). Cohen's d shows a medium effect size for both estimates (both $d = 0.52$).

5. Conclusions

The results support that participation in the 'Sustainable Development' course induced positive changes in the students' environmental citizenship. The presentation at iREEC2022 will also highlight the specific pedagogical design of the 'Sustainable Development' course and relate it to the learning outcomes that we observed in our study. As such, our paper provides a timely contribution, shedding light on how specific pedagogical approaches in higher education can foster environmental citizenship.

Author Contributions: Conceptualization, J.B.-d.P., A.T., D.G. and R.H.; methodology, J.B.-d.P., A.T., D.G. and R.H; software, J.B.-d.P., A.T., D.G. and R.H; validation, J.B.-d.P., A.T., D.G. and R.H.; formal analysis, J.B.-d.P., A.T., D.G. and R.H; investigation, J.B.-d.P., A.T., D.G. and R.H; resources, J.B.-d.P., A.T., D.G. and R.H; data curation, J.B.-d.P., A.T. and R.H; writing—original draft preparation, J.B.-d.P., A.T., D.G. and R.H; writing—review and editing, J.B.-d.P., A.T., D.G. and R.H; visualization, J.B.-d.P., A.T., D.G. and R.H; supervision, J.B.-d.P., A.T., D.G. and R.H; project administration, A.T.; funding acquisition, J.B.-d.P., A.T., D.G. and R.H. All authors have read and agreed to the published version of the manuscript.

Funding: This research received no external funding.

Institutional Review Board Statement: The study was conducted according to the guidelines of the Declaration of Helsinki, and approved by the Institute of Social Sciences, Arts and Humanities at Kaunas University of Technology (protocol no. V19-1253-7-1, 27 August 2020).

Informed Consent Statement: Informed consent was obtained from all subjects involved in the study.

Data Availability Statement: A publicly archived dataset with a questionnaire is available through LiDA: https://hdl.handle.net/21.12137/RZJ0FL (accessed on 22 December 2021).

Acknowledgments: This study is partly based on work from Cost Action ENEC (European Network for Environmental Citizenship) (CA16229), supported by COST (European Cooperation in Science and Technology). The authors also want to thank the lecturers and the students of the "Sustainable Development" and the "Media Philosophy" courses at Kaunas University of Technology, for supporting or taking part in the investigation.

Conflicts of Interest: The authors declare no conflict of interest.

References

1. Hadjichambis, A.C.; Reis, P.; Paraskeva-Hadjichambi, D.; Činčera, J.; Boeve-de Pauw, J.; Gericke, N.; Knippels, M.-C. (Eds.) *Conceptualizing Environmental Citizenship for 21st Century Education; Environmental Discourses in Science Education*; Springer International Publishing: Cham, Switzerland, 2020; p. 261. ISBN 978-3-030-20248-4.
2. Hadjichambis, A.C.; Paraskeva-Hadjichambi, D. Environmental Citizenship Questionnaire (ECQ): The Development and Validation of an Evaluation Instrument for Secondary School Students. *Sustainability* **2020**, *12*, 821. [CrossRef]
3. Telešienė, A.; Boeve-de Pauw, J.; Goldman, D.; Hansmann, R. Evaluating an Educational Intervention Designed to Foster Environmental Citizenship among Undergraduate University Students. *Sustainability* **2021**, *13*, 8219. [CrossRef]

environmental sciences proceedings

MDPI

Proceeding Paper

European Green Deal and Environmental Citizenship: Two Interrelated Concepts †

Andreas Ch. Hadjichambis [1,2]

1 Cyprus Centre for Environmental Research & Education (CYCERE), Limassol 3304, Cyprus; a.hadjichambis@cytanet.com.cy; Tel.: +357-9947-7309

2 Cyprus Ministry of Education, Culture, Sport & Youth (MOEC), Nicosia 1434, Cyprus

† Presented at the 2nd International Conference of International Researchers of the Education for Environmental Citizenship 2022, 10–11 March 2022. Available online: https://enec-cost.eu/ireec22/.

Abstract: The world is facing an unprecedented global environmental crisis as environmental problems have been exacerbated in recent decades. Climate crisis, plastic pollution, and the loss of biodiversity are just some of the many environmental issues we are facing every day. Actions by citizens are central to EU plans to tackle the recent environmental crisis, to achieve the European Green Deal, and the EU 2050 strategy for a low (neutral) carbon Europe. Perhaps, more than any other previous environmental policy, the European Green Deal (EGD) has set participation and citizen engagement as one of its main priorities. Empowering citizens for transition towards a climate neutral, sustainable Europe is one of the horizontal priority areas of the EGD (Thematic area 10: Empowering citizens for transition towards a climate neutral, sustainable Europe Call). According to the EGD, the green transition must be just and inclusive and requires ambitious actions to engage people, communities, and organizations to bring about a fair and inclusive transition, leaving no-one behind. Such actions must promote change at the collective level through deliberation, as well as through research to foster behavioral and social change, and at an individual level by empowering citizens as "agents of change". This is a fundamental aim of the recent conceptualization of Environmental Citizenship.

Keywords: European Green Deal; environmental citizenship; citizen engagement; green transformation; Education for Environmental Citizenship (EEC)

Citation: Hadjichambis, A.C. European Green Deal and Environmental Citizenship: Two Interrelated Concepts. *Environ. Sci. Proc.* **2022**, *14*, 3. https://doi.org/10.3390/environsciproc2022014003

Academic Editors: Pedro Reis, Marie-Christine Knippels, Audronė Telesiene, Daphne Goldman, Demetra Paraskeva-Hadjichambi, Jan Cincera and Kateřina Jančaříková

Published: 1 March 2022

1. Introduction

The recent conceptualization of Environmental Citizenship by the European Network for Environmental Citizenship [1] is a very important research development. Based on the EEC Model (see Figure 1), ENEC developed a consensus on what Environmental Citizenship is, on how Environmental Citizenship can be effective for Environmental Policy, and what Environmental Citizenship should include for the adequate empowerment of citizens for transition towards a climate neutral, sustainable Europe.

Due to the very important relationship between Environmental Citizenship and the EGD, the EU Directorate for Research and Innovation together with the COST Association organized a special workshop to determine the research priorities in future research calls.

Figure 1. The EEC Model [Source: Hadjichambis & Paraskeva-Hadjichambi, (2020)] [2].

2. Methodology

This study theoretically examines the relationships between two very important concepts in the recent environmental sciences area: the European Green Deal (EGD) and Environmental Citizenship (EC).

3. Findings

One of the most critical issues that encroaches on future citizens is the current fragmentation of knowledge production and training systems [3]. Particularly, in the last two decades, the separation of knowledge and expertise has been identified as a problem in environmental science, policy-making, and education in general [4]. Environmental education remains highly specialized and predominantly focused on natural sciences, while citizenship, economic and social issues, environmental justice aspects, and structural causes of environmental problems are only marginally included in environmental education programs, if not at all [5]. It remains entirely unclear how environmental citizenship matures in universities and schools, in ways that promote innovation, sustainability and green entrepreneurship, and growth. It is unknown how future citizens will acquire the needed knowledge, competencies, skills, values, and behaviors to contribute meaningfully to a positive green future. This means that these aspects are central in stimulating environmental citizenship in future environmental citizens, who will then be equipped with the ways of thinking and acting that promote and generate a green sustainable future.

For Europe, becoming the world's first climate-neutral continent by 2050 is a once in a lifetime opportunity to modernize the EU's economy and society and re-orient them towards a just and sustainable future. Research and innovation will play a central role in (a) accelerating and navigating the necessary transitions; (b) deploying, demonstrating, and de-risking solutions; and (c) engaging citizens in social innovation [6]. Horizontal area 10 is dedicated to empowering citizens and has three sub-topics: (a) European research infrastructures' capacities and services to address EGD challenges; (b) Behavioral, social, and cultural change for EGD; and (c) Enabling citizens to act on climate change and environmental protection through education, citizen science, observation initiatives, and civic involvement. Environmental Citizenship (EC) has a central role in behavioral, social, and cultural change as EC aims to achieve critical and active engagement and civic participation of the citizens, as well as in practicing environmental rights and duties, and promoting inter- and intra-generational justice [1]. Education for environmental citizenship (EEC) is a channel through which citizens can be enabled to act on climate change and environmental protection.

Five green mission areas in Horizon Europe directly support the European Green Deal: (1) Restore our ocean and waters, (2) Climate-neutral and smart cities, (3) A soil deal for Europe, (4) Adaptation to climate change and societal transformation.

As more people on the planet live in cities, it is necessary for Environmental Citizenship to be practiced in cities in order to address urban environmental problems and promote urban sustainability. Therefore, green and climate-resilient cities can be a case study to examine the relationships between EGD and EC.

Environmental Citizenship in Green and Sustainable cities focuses on citizens as social change agents. Such citizens not only engage in sustainable household practices but respect the importance of raising awareness, discussion, and debates on sustainability policies for the common good and maintenance of the Earth's ecosystems [7]. Decision Support Systems for citizens' active participation and engagement for improved management of cities and society is a very important issue. Social aspects, impacts and resilience of cities, citizen behavior monitoring, analysis and change within urban communities, environmental governance, and environmental justice can be proved to be crucial dimensions for green and sustainable cities. Furthermore, Citizen Science—which relies on citizen participation for continuous knowledge generation (e.g., [8], knowledge transfers over time [9], and practical environmental management—is of outstanding importance in achieving climate resilience in contemporary cities. Finally, Nature-based Solutions (NbS), by using ecosystem services, are innovative solutions that use natural elements to achieve environmental and societal goals in sustainable cities [10], with the effective participation of environmental citizens as an internal prerequisite.

The active and catalytic role of citizens and their direct civic participation is essential to address climate change and other environmental problems [11]. Changes in citizens' and consumers' behaviors towards more sustainable patterns can happen through education [12], especially Education for Environmental Citizenship as it is defined by ENEC [1]. This novel type of education can provide citizens with the knowledge, values, attitudes, skills, competencies, and behaviors required for monitoring their environmental impacts, and for deep civic participation in individual and collective spheres and in private and public spheres. According to the European Green Deal Call, "It is crucial to directly involve citizens and communities in contributing to climate action and protecting the environment, thereby encouraging them to change their personal behaviour, reducing their carbon and environmental footprint and taking action at the individual and collective level. This would lead to a more sustainable lifestyle and relationship to the environment, by promoting biodiversity protection, nature-based solutions for climate resilience, sustainable energy consumption, waste management, etc." [13] (p. 1). These integral parameters could play a role in the Circular Economy and Cities with Climate Resilience.

4. Conclusions

This study helps to determine how current and prospective dimensions of the concept of "environmental citizenship" are related to the "European Green Deal" initiative. Citizens play a critical role in climate action and protecting the environment, thereby encouraging them to change their personal behavior, reducing their carbon and environmental footprint, and taking action at the individual and collective level; thus, the concepts of Environmental Citizenship and the European Green Deal are inseparable. This connection would lead to a more sustainable lifestyle and relationship of citizens to nature and the environment.

Funding: This research received no external funding.

Institutional Review Board Statement: Not applicable.

Informed Consent Statement: Not applicable.

Data Availability Statement: No new data were created or analyzed in this study. Data sharing is not applicable to this article.

Acknowledgments: This study is partly inspired by Cost Action ENEC—European Network for Environmental Citizenship (CA16229), supported by COST (European Cooperation in Science and Technology).

Conflicts of Interest: The authors declare no conflict of interest.

References

1. European Network for Environmental Citizenship–ENEC. Defining "Environmental Citizen". 2018. Available online: http://enec-cost.eu/environmental-citizen/ (accessed on 2 February 2022).
2. Hadjichambis, A.C.; Paraskeva-Hadjichambi, D. Education for environmental citizenship: The pedagogical approach. In *Conceptualizing Environmental Citizenship for 21st Century Education*; Environmental Discourses in Science Education; Hadjichambis, A.C., Reis, P., Paraskeva-Hadjichambi, D., Činčera, J., Boeve-de Pauw, J., Gericke, N., Knippels, M.-C., Eds.; Springer: Cham, Switzerland, 2020; Volume 4, pp. 237–261.
3. Rose, D.C.; Mukherjee, N.; Simmons, B.I.; Tew, E.R.; Robertson, R.J.; Vadrot, A.B.; Doubleday, R.; Sutherland, W.J.Y. Policy windows for the environment: Tips for improving the uptake of scientific knowledge. *Environ. Sci. Policy* **2020**, *113*, 47–54. [CrossRef]
4. Twalo, T. Challenges of knowledge production and knowledge use among researchers and policy-makers. *Educ. Action Res.* **2019**, *27*, 269–285. [CrossRef]
5. Hadjichambis, A.C.; Reis, P.; Paraskeva-Hadjichambi, D.; Činčera, J.; Boeve-de Pauw, J.; Gericke, N.; Knippels, M.-C. (Eds.) *Conceptualizing Environmental Citizenship for 21st Century Education*; Environmental Discourses in Science Education; Springer: Cham, Switzerland, 2020.
6. European Commission. Research and Innovation for the European Green Deal. 2022. Available online: https://ec.europa.eu/info/research-and-innovation/strategy/strategy-2020-2024/environment-and-climate/european-green-deal_en (accessed on 2 February 2022).
7. Cook, N. *Sustainability Citizenship in Cities: Theory and Practice*; Routledge (Earthscan): London, UK, 2016.
8. Krasny, M.E.; Russ, A.; Tidball, K.G.; Elmqvist, T. Civic ecology practices: Participatory approaches to generating and measuring ecosystem services in cities. *Ecosyst. Serv.* **2014**, *7*, 177–186. [CrossRef]
9. Andersson, E.; Barthel, S. Memory carriers and stewardship of metropolitan landscapes. *Ecol. Indic.* **2016**, *70*, 606–614. [CrossRef]
10. Kabisch, N.; Korn, H.; Stadler, J.; Bonn, A. *Nature-Based Solutions to Climate Change Adaptation in Urban Areas: Linkages between Science, Policy and Practice*; Springer International Publishing: Cham, Switzerland, 2017.
11. Simonova, P.; Cincera, J.; Kroufek, R.; Krepelkova, S.; Hadjichambis, A.C. Active citizens: Evaluation of a community-based education program. *Sustainability* **2019**, *11*, 663.
12. Paraskeva-Hadjichambi, D.; Goldman, D.; Hadjichambis, A.C.; Parra, G.; Lapin, K.; Knippels, M.C.; Van Dam, F. Educating for environmental citizenship in non-formal frameworks for secondary level youth. In *Conceptualizing Environmental Citizenship for 21st Century Education*; Environmental Discourses in Science Education, Hadjichambis, A.C., Reis, P., Paraskeva-Hadjichambi, D., Činčera, J., Boeve-de Pauw, J., Gericke, N., Knippels, M.-C., Eds.; Springer: Cham, Switzerland, 2020; Volume 4, pp. 213–235.
13. European Commission. Enabling Citizens to Act on Climate Change and Environmental Protection through Education, Citizen Science, Observation Initiatives, and Civic Involvement. 2021. Available online: https://ec.europa.eu/info/sites/default/files/research_and_innovation/green_deal/gdc_stakeholder_engagement_topic_10-3_civic_involvement.pdf (accessed on 12 February 2022).

environmental sciences proceedings

MDPI

Abstract

Environmental Citizenship of Dutch Lower Secondary Students †

Michiel van Harskamp *, Marie-Christine P. J. Knippels and Wouter R. van Joolingen

Freudenthal Institute, Utrecht University, P.O. Box 85.170, 3508 AD Utrecht, The Netherlands; m.c.p.j.knippels@uu.nl (M.-C.P.J.K.); w.r.vanjoolingen@uu.nl (W.R.v.J.)

* Correspondence: m.vanharskamp@uu.nl; Tel.: +31-30-253-8251

† Presented at the 2nd International Conference of International Researchers of the Education for Environmental Citizenship 2022, 10–11 March 2022. Available online: https://enec-cost.eu/ireec22/.

Keywords: environmental citizenship; lower secondary level; students; interviews

Citation: van Harskamp, M.; Knippels, M.-C.P.J.; van Joolingen, W.R. Environmental Citizenship of Dutch Lower Secondary Students. *Environ. Sci. Proc.* **2022**, *14*, 2. https://doi.org/10.3390/environsciproc2022014002

Academic Editors: Andreas Ch. Hadjichambis, Pedro Reis, Audronė Telesiene, Daphne Goldman, Demetra Paraskeva-Hadjichambi, Jan Cincera and Kateřina Jančaříková

Published: 1 March 2022

1. Rationale

To enable students to deal with sustainability issues, science education needs to provide tools to navigate sustainability issues, thus fostering Environmental Citizenship (EC). However, stimulating EC through science education remains challenging for science teachers. For instance, they experience challenges with guiding dialogue, and deem the curricula unsuitable [1]. Understanding current levels of EC among students would help science teachers in developing and teaching effective EC lessons. Currently, most of what we know about student EC is derived from large scale, quantitative studies. A more in-depth, qualitative view on student EC is desirable [2].

2. Key Objectives

In order to better support science teachers during education for EC, this study provides a qualitative view on the current level of EC of Dutch lower secondary students. The research question is: How can Dutch lower secondary students' EC be characterized? This characterization will be based on EC knowledge, attitudes, behaviours, and reflection, together forming student EC competence.

3. Research Design and Methodology

We conducted semi-structured interviews with 42 students (F: 25, M: 17; average age 13.3). Participants were interviewed at their own schools, which were selected to ensure geographical spread representing the urbanization level of the Netherlands (90.5% urban, 9.5% rural). Interviews lasted 11 minutes on average. The interview questions related to students' EC knowledge, attitudes, behaviour, and reflection. Interviews were audio recorded and transcribed verbatim. Transcripts were coded using open coding, and interrater agreement was calculated for 70% of the coded data (with ultimately 86.2% agreement between researchers, before discussion of coded data).

4. Findings and Conclusions

The data show that lower secondary students in the Netherlands have relatively narrow views of sustainability. They mainly relate it to sustainable energy use and recycling. About 90% of the students think about the future of the planet at least sometimes, with over half of them being worried. They fear for doom scenarios with mass extinction, hunger, pollution, and destruction of the planet. However, about half of the students notice sustainability issues close to home, whereas almost all of them think sustainability issues exist elsewhere. This means students feel distant in regard to sustainability issues. In contrast, students are interested in learning more about possible solutions that they can implement themselves. Students do not discuss sustainability with friends, but about half of them discuss them at home or at school. They often adopt a form of EC that can be typified as personally responsible citizenship [3].

Author Contributions: Conceptualization, M.v.H., M.-C.P.J.K. and W.R.v.J.; methodology, M.v.H., M.-C.P.J.K. and W.R.v.J.; analysis, M.v.H. and M.-C.P.J.K.; writing—original draft preparation, M.v.H.; writing—review and editing, M.v.H., M.-C.P.J.K. and W.R.v.J.; visualization, M.v.H.; funding acquisition, M.v.H., M.-C.P.J.K. and W.R.v.J. All authors have read and agreed to the published version of the manuscript.

Funding: This work was supported by the NRO (Nationaal Regieorgaan Onderwijsonderzoek) under Grant number 40.5.18540.030.

Institutional Review Board Statement: This study was performed under ethical guidelines from the Faculty of Science from Utrecht University; explicit ethical review for this study was not mandatory.

Informed Consent Statement: Informed consent was obtained for all subjects involved in the study.

Data Availability Statement: Data sharing not applicable.

Conflicts of Interest: The authors declare no conflict of interest.

References

1. Van Harskamp, M.; Knippels, M.C.P.J.; Van Joolingen, W.R. Secondary Science Teachers' Views on Environmental Citizenship in The Netherlands. *Sustainability* **2021**, *13*, 7963. [CrossRef]
2. Blatt, E. Uncovering students environmental identity: An exploration of activities in an environmental science course. *J. Environ. Educ.* **2014**, *45*, 194–216. [CrossRef]
3. Westheimer, J. On the Relationship Between Political and Moral Engagement. In *Getting Involved: Global Citizenship Development and Sources of Moral Values*; Oser, F.M., Veugelers, W., Eds.; Sense Publishers: Rotterdam, The Netherlands, 2008; pp. 17–29.

environmental sciences proceedings

MDPI

Abstract

The Role of Education, Self—Reported Knowledge and Environmental Risk Perception in Disaster Preparedness †

Aistė Balžekienė

Civil Society and Sustainability Research Group, Kaunas University of Technology, 44249 Kaunas, Lithuania; aiste.balzekiene@ktu.lt

† Presented at the 2nd International Conference of International Researchers of the Education for Environmental Citizenship 2022, 10–11 March 2022. Available online: https://enec-cost.eu/ireec22/.

Abstract: One of the dimensions of environmental citizenship is related to the understanding of the structural causes of environmental degradation and it also emphasizes the empowerment of citizens. With increasing threats from climate change impacts, the United Nations emphasizes the importance of local capacities for environmental emergency preparedness. Research shows that education plays an important role in the level of disaster preparedness (for example, refs. [1,2]); moreover, environmental knowledge is an important factor for shaping public-sphere environmental behaviors (for example, ref. [3]). The aim of this presentation is to explore to what degree the level of education, environmental knowledge and risk perception influences the level of preparedness for environmental emergencies in Lithuania. Lithuania is one of the countries where climate change impacts have drastically increased during recent years; thus, it is important to research the role and preparedness of citizens to cope with related threats and to indicate knowledge gaps in the understanding of causal relations. This presentation uses data from two datasets, which are collected during the Risk-Space project in Lithuania: (1) the representative survey of Lithuanian population, conducted in 2020, with 2007 respondents in Lithuania; and (2) experts from municipalities in Lithuania, responsible for civil contingency. Conducted in 2021. Experts number 58 (out of 60) participated in the study. The items that are analysed in this presentation include the following: Independent variables include the perception of specific environmental risks (floods, forest fires, air pollution and water pollution), self-reported knowledge about environmental threats, and the level of education of the respondent; dependent variables include the level of preparedness of individual and municipality to cope with environmental emergencies. Results reveal that the self-reported level of disaster preparedness is critically low both evaluating the preparedness of the municipality where the respondent lives and the individual preparedness. The level of education and self-reported environmental knowledge is significantly positively correlated with the individual's disaster preparedness level; however, they are not correlated with the evaluation of municipality preparedness.

Keywords: environmental emergencies; environmental knowledge; environmental attitudes; disaster preparedness

Citation: Balžekienė, A. The Role of Education, Self—Reported Knowledge and Environmental Risk Perception in Disaster Preparedness. *Environ. Sci. Proc.* **2022**, *14*, 5. https://doi.org/10.3390/environsciproc2022014005

Academic Editors: Andreas Ch. Hadjichambis, Pedro Reis, Marie-Christine Knippels, Audronė Telesiene, Daphne Goldman, Demetra Paraskeva-Hadjichambi, Jan Cincera and Kateřina Jančaříková

Published: 1 March 2022

Funding: Presentation is based on research project "Mapping of Risk Perception in Lithuania: Spatial and Socio-psychological Dimensions" (Risk-Space), funded by Research Council of Lithuania (S-MIP-19-28).

Informed Consent Statement: Informed consent was obtained from all subjects involved in the study.

Data Availability Statement: The data presented in this study are openly available in the Dataverse repository of the Lithuanian Data Archive for HSS: https://hdl.handle.net/21.12137/Q0YILI, (accessed on 22 February 2022).

Conflicts of Interest: The author declares no conflict of interest.

References

1. Xu, D.; Peng, L.; Liu, S.; Wang, X. Influences of Risk Perception and Sense of Place on Landslide Disaster Preparedness in Southwestern China. *Int. J. Disaster Risk Sci.* **2018**, *9*, 167–180. [CrossRef]
2. Domingues, R.B.; Santos, M.C.; de Jesus, S.N.; Ferreira, Ó. How a coastal community looks at coastal hazards and risks in a vulnerable barrier island system (Faro Beach, southern Portugal). *Ocean Coast. Manag.* **2018**, *157*, 248–256. [CrossRef]
3. Hansmann, R.; Binder, C.R. Determinants of different types of positive environmental behaviors: An analysis of public and private sphere actions. *Sustainability* **2020**, *12*, 8547. [CrossRef]

environmental sciences proceedings

MDPI

Abstract

Environmental Citizenship or Consumership—A Methodological Approach †

Vladislav Kaputa *, Hubert Paluš, Hana Maťová and Mikuláš Šupín

Department of Marketing, Trade and World Forestry, Technical University in Zvolen, 96001 Zvolen, Slovakia; palus@tuzvo.sk (H.P.); matova@tuzvo.sk (H.M.); supin@tuzvo.sk (M.Š.)

* Correspondence: kaputa@tuzvo.sk; Tel.: +421-455-206-590

† Presented at the 2nd International Conference of International Researchers of the Education for Environmental Citizenship 2022, 10–11 March 2022. Available online: https://enec-cost.eu/ireec22/.

Abstract: This work deals with the proposal of a methodology for the survey of citizens' attitudes in relation to their citizenship in light of the economic dimensions of environmental citizenship (EC). This is especially beneficial for university students, doctoral students and young researchers. The design allows for quantitative data processing. Thus, closed questions are dominant. However, open questions may serve as an incentive to set up further research. Demographic questions (age, education, gender, residence, social status, etc.) are used to perform a contingency analysis. Relevant hypotheses will be tested with suitable statistical methods, depending on the method of sampling (random vs. intentional). How the respondents perceived the status of being a citizen shall be investigated using the semantic differential method—a pair of opposite adjectives on the value scale. The proposed pairs are: actual–outdated, active–passive, easy–hard, significant–meaningless, normal–subnormal, attractive–repulsive. The concepts of a global citizen [1,2] and "Environmental Citizen" defined by ENEC [3] point to the citizen as a mediator of change from the local to global level. Thus follows, especially within EU countries, the verification of whether citizenship is perceived as more "regional" (country affiliation) or more as "European". This attitude can be further related to attitudes towards EU institutions. Based on the above definitions, we propose determining the involvement of respondents in civic activities (social and environmental level, etc.). The frequency of involvement is examined at set levels from "none" (through sporadic) to "permanent". Barry [4] criticized firms and public bodies for adopting the language of EC as being motivated either by compliance with corporate environmental reporting or as evidence of a commitment to the concept of CSR. Here, encouraging employees to be environmental citizens is simply an integral part of corporate systems. Barry describes such EC as a part-time occupation—something one engages in during working hours [5]. Thus, we include a question focused on the transfer of habits and standards of behavior from the organization to civic life (the answers use a Likert-type scale). The potential transfer of habits can be at the following levels: ethical principles, responsibility for the environment, the ways of communicating with people, and the relationship with local communities. Moreover, the cultural environment specifics could be included. The next part focuses on which entities play an important role in environmental responsibility: the state (institutions); EU institutions; businesses; individuals (their civic activities); individuals (their consumer decisions). Considering the last two items, the awareness of citizenship is surveyed with regard to: having rights, having responsibility, and having an affiliation with society. From the consumer's position, consumer rights, responsibility (for purchasing decisions), and belonging to a group of consumers are also examined. The respondents assess the extent to which their purchasing behavior is affected by: global trends, their own discretion, personalities or institutions, media, and local influences.

Keywords: environmental citizenship; citizen; consumer; methodology; questionnaire

Citation: Kaputa, V.; Paluš, H.; Maťová, H.; Šupín, M. Environmental Citizenship or Consumership—A Methodological Approach. *Environ. Sci. Proc.* **2022**, *14*, 7. https://doi.org/10.3390/environsciproc2022014007

Academic Editors: Andreas Ch. Hadjichambis, Pedro Reis, Marie-Christine Knippels, Audronė Telesiene, Daphne Goldman, Demetra Paraskeva-Hadjichambi, Jan Cincera and Kateřina Jančaříková

Published: 2 March 2022

Author Contributions: Conceptualization, V.K. and H.P.; methodology, V.K. and H.M.; writing—original draft preparation, V.K. and M.Š.; writing—review and editing, supervision, and project administration, V.K. and H.P. All authors have read and agreed to the published version of the manuscript.

Institutional Review Board Statement: Not applicable.

Informed Consent Statement: Not applicable.

Data Availability Statement: Not applicable.

Acknowledgments: The authors were inspired by the ideas arising from the project COST Action CA16229–European Network for Environmental Citizenship. The authors are also grateful for the support of the Scientific Grant Agency of the Ministry of Education, Science, Research and Sport of the Slovak Republic and the Slovak Academy of Sciences, Grant No. 1/0475/22 "Environmental consumer and environmental citizen".

Conflicts of Interest: The authors declare no conflict of interest.

References

1. Reysen, S.; Larey, L.W.; Katzarska-Miller, I. College Course Curriculum and Global Citizenship. *Int. J. Develop. Educ. Global Learn.* **2012**, *4*, 27–40. [CrossRef]
2. Rosa, W.; Shaw, H.K. *Global Nurse Citizenship: Toward a Safe and Inclusive Civil Society*; Springer: New York, NY, USA, 2017. [CrossRef]
3. European Network for Environmental Citizenship—ENEC (2018). Defining "Environmental Citizen". Available online: http://enec-cost.eu/environmental-citizen/ (accessed on 30 September 2021).
4. Barry, J. Resistance is Fertile: From Environmental to Sustainability Citizenship. In *Environmental Citizenship*; Dobson, A., Bell, D., Eds.; MIT Press: Cambridge, MA, USA, 2006; pp. 21–48.
5. Kaputa, V.; Lapin, K.; Leregger, F.; Gekic, H. Economic dimensions of environmental citizenship. In *Conceptualizing Environmental Citizenship for 21st Century Education*; Hadjichambis, A.C., Reis, P., Hadjichambi, D.P., Činčera, J., Boeve-de Pauw, J., Gericke, N., Knippels, M.C., Eds.; Springer: Cham, Switzerland, 2020; pp. 29–48, ISBN 978-3-030-20248-4.

environmental sciences proceedings

MDPI

Abstract

Concepts of Empowerment: The Role of Energy and Environmental Citizenship in the Energy Transition †

Mehmet Efe Biresselioglu [1], Giuseppe Carrus [2], Frances Fahy [3], Andrea Kollmann [4,*], Jens Olgard Dalseth Røyrvik [5] and David C. Finger [4,6]

1 Sustainable Energy Division, Izmir University of Economics, 35330 Izmir, Turkey; efe.biresselioglu@ieu.edu.tr
2 Department of Educational Sciences, Università Roma Tre, 00146 Rome, Italy; giuseppe.carrus@uniroma3.it
3 Geography, Archaeology & Irish Studies, National University of Ireland, H91 TK33 Galway, Ireland; frances.fahy@nuigalway.ie
4 Energy Engineering Department, Energieinstitut an der Johannes Kepler Universität, 4040 Linz, Austria; finger@energieinstitut-linz.at
5 Departement of Social Anthropology, Norwegian University of Science and Technology, 7491 Trondheim, Norway; jens.royrvik@samforsk.no
6 School of Engineering, Reykjavik University, 102 Reykjavik, Iceland
* Correspondence: kollmann@energieinstitut-linz.at; Tel.: +43-732-2468-5660
† Presented at the 2nd International Conference of International Researchers of the Education for Environmental Citizenship 2022, 10–11 March 2022. Available online: https://enec-cost.eu/ireec22/.

Abstract: This symposium brings together social science energy researchers working on gaining an understanding of the various social, economic and political dimensions of energy citizenship and how energy citizenship affects the clean-energy transition process across Europe. Representing six different countries (Austria, Iceland, Ireland, Italy, Norway and Turkey) the contributors each present cutting-edge scientific research activities that exemplify developments toward achieving energy citizenship and reflect on the overlaps and synergies with environmental citizenship.

Keywords: energy citizenship; environmental citizenship; energy transition

Citation: Biresselioglu, M.E.; Carrus, G.; Fahy, F.; Kollmann, A.; Røyrvik, J.O.D.; Finger, D.C. Concepts of Empowerment: The Role of Energy and Environmental Citizenship in the Energy Transition. *Environ. Sci. Proc.* **2022**, *14*, 9. https://doi.org/10.3390/environsciproc2022014009

Academic Editors: Andreas Ch. Hadjichambis, Pedro Reis, Marie-Christine Knippels, Audronė Telesiene, Daphne Goldman, Demetra Paraskeva-Hadjichambi, Jan Cincera and Kateřina Jančaříková

Published: 3 March 2022

Publisher's Note: MDPI stays neutral with regard to jurisdictional claims in published maps and institutional affiliations.

1. Introduction

The IPCC 2021 report emphasized the urgency to reduce global greenhouse gas emissions (GHG) in order to avoid the catastrophic impacts of global warming surpassing 1.5 °C. In order to achieve the goals outlined in this report, a sustainable transition process towards climate neutrality needs to be developed for the energy sector. In this symposium, we intend to discuss a core aspect of this process: the role, rights, duties and opportunities of citizens in this transition. This symposium aims at presenting social, economic and political dimensions of energy citizenship and how energy citizenship affects the clean-energy transition process across Europe. Representing six different countries (Austria, Iceland, Ireland, Italy, Norway and Turkey) the contributors present cutting-edge scientific research activities in the field of energy citizenship and reflect on the overlaps and synergies with environmental citizenship.

Mehmet Efe Biresselioglu discusses the relationship of community and individual dynamics in the context of energy citizenship, along with main themes, dimensions, indicators and variables associated with energy citizenship and how these act as drivers towards energy citizenship, based on evidence from the Literature Review of the Horizon 2020 DIALOGUES Project.

Giuseppe Carrus presents on the psychological drivers of sustainable energy choices and pro-environmental behaviors among individuals, groups and communities, focusing on the interplay between individual and collective factors to promote energy citizenship, and will discuss theories and empirical findings of the DIALOGUES project and previous EU-funded research projects.

Andrea Kollmann presents on the Horizon 2020 project DIALOGUES, which aims at elaborating the knowledge needed to harness the concept of energy citizenship for targeted policymaking to strengthen the role and efficacy of energy citizenship as a contributor to the strategic priorities of the sustainable energy transition.

Frances Fahy presents on the European EnergyPROSPECTS project which aims to disclose the societal conditions conducive to the thriving of engaged, sustainability-oriented, democratic or otherwise desirable forms of energy citizenship.

Jens Olgard Dalseth Røyrvik presents how the duties and rights (i.e., citizenship) related to energy manifests differently in urban and rural areas in Norway, and further how this creates both active participation and active resistance to the green shift.

The symposium is moderated by David C. Finger. Educators, practitioners and academics are invited to attend this panel session and discuss recent developments in the field of energy citizenship.

2. Structure of the Symposium

The symposium will be introduced by David C. Finger, who also moderates the panel discussion following the presentations of the five speakers:

- Mehmet Efe Biresselioglu: Community and Individual Dynamics of Energy Citizenship: Evidence from the Literature Review of the DIALOGUES Project
- Giuseppe Carrus: Individual and Collective Factors at the Basis of pro-Environmental Behaviors, Sustainable Energy Choices and Energy Citizenship
- Frances Fahy: EnergyPROSPECTS–Current Perspectives on Energy Citizenship
- Andrea Kollmann: Conceptualizing Energy Citizenship–Insights from the DIALOGUES Project
- Jens Olgard Dalseth Røyrvik: The Difference of Urban and Rural Energy Citizenship in Norway

Funding: The research by Biresselioglu, M.E., Carrus, G., Kollmann, A., Røyrvik, J. is funded by the European Union's Horizon 2020 research and innovation program under grant agreement No 101022585. The research by Fahy, F. is funded by the European Union's Horizon 2020 research and innovation program under grant agreement No 101022492.

Institutional Review Board Statement: Not applicable.

Informed Consent Statement: Informed consent was obtained from all subjects involved in the study.

Conflicts of Interest: The authors declare no conflict of interest.

environmental sciences proceedings

MDPI

Proceeding Paper

Investigating the Participation Facets of Environmental Citizen Science Initiatives: A Systematic Literature Review of Empirical Research[†]

Michalis A. Vasiliades [1], Andreas Ch. Hadjichambis [1,2,*], Demetra Hadjichambi [1,2], Anastasia Adamou [1] and Yiannis Georgiou [1,2]

1 Cyprus Centre for Environmental Research and Education (CYCERE), Limassol 3304, Cyprus; mv.kykpee@gmail.com (M.A.V.); d.hadjichambi@cytanet.com.cy (D.H.); na.kykpee@gmail.com (A.A.); yg.kykpee@gmail.com (Y.G.)
2 Cyprus Ministry of Education, Culture, Sport and Youth (MOEC), Nicosia 1434, Cyprus
* Correspondence: a.hadjichambis@cytanet.com.cy; Tel.: +357-99477309

† Presented at the 2nd International Conference of International Researchers of the Education for Environmental Citizenship 2022, 10–11 March 2022. Available online: https://enec-cost.eu/ireec22/.

Abstract: Citizen science (CS) has shown tremendous popularity in recent years; however, there is still a lack of understanding of important aspects that determine citizens' participation and involvement in CS initiatives. Although CS initiatives could serve as a means of promoting forms of participation that contribute to the democratization of science, limited attention is still being paid to the "citizen" component of the citizen science term. For this reason, a systematic literature review (SLR), aligned with the PRISMA methodology, was applied to empirical studies on citizens' participation in environmental and nature-based CS initiatives established over the last two decades. The participatory aspect of the retrieved 119 CS initiatives was analysed on the basis of: (a) citizens' participation and (b) environmental citizenship. Our findings show that the majority of the CS initiatives were mostly limited to the local scale, and they primarily followed the contributory model, in which volunteers were mostly treated as "data collectors". Therefore, it is important to overcome barriers related to the design and implementation of CS that hinder citizens' participation and, at the same time, to strengthen democratization through a more participatory engagement of active and aware citizens, thus promoting environmental citizenship.

Keywords: participation aspects; environmental citizen science; facilitators; constraints; models and practices; education for environmental citizenship

Citation: Vasiliades, M.A.; Hadjichambis, A.C.; Hadjichambi, D.; Adamou, A.; Georgiou, Y. Investigating the Participation Facets of Environmental Citizen Science Initiatives: A Systematic Literature Review of Empirical Research. *Environ. Sci. Proc.* **2022**, *14*, 1. https://10.3390/environsciproc2022014001

Academic Editors: Pedro Reis, Marie-Christine Knippels, Audronė Telesiene, Daphne Goldman, Jan Cincera and Kateřina Jančaříková

Published: 25 February 2022

1. Introduction

Citizen science (CS) has a long history, and it describes the involvement of members of the public in scientific research, mainly through forms of data collection and information gathering in various fields of science, such as nature, ecology, and the environment [1–5]. The interaction between the public and scientists can be considered as a two-way process that, on the one hand, promotes public participation in science, where among other things, the former acquire knowledge and skills during their involvement in data collection processes, and on the other hand, the latter use the large-scale data collected, which may not be easily gathered with other techniques, for the common good [6,7]. Despite the long history of CS, there are still several limitations in the underlying models and frameworks that guide public participation in CS initiatives. In many CS initiatives, the citizens are merely treated as "data collectors" or "passive" participants [8], instead of "active citizens" [9]. This finding is not surprising, given that CS research is not focused, for instance, on how CS can help the public to understand and appreciate the power of science for socio-political and socio-cultural action [4].

However, CS can be an excellent venue for the democratization of science, by achieving an "inclusive" and active engagement of citizens in the scientific process. This study presents a systematic literature review (SLR) of the available empirical research on citizens' participation in environmental and nature-based CS. More specifically, the participatory aspects of a total of 119 CS initiatives were analysed according to: (a) the CS models and practices defining citizens' participation, and (b) the participation forms for the achievement of education for environmental citizenship.

2. Methodology

The published literature was surveyed using four electronic databases (Scopus, Web of Science, Education Research Complete, ERIC), and the retrieval of the empirical studies followed three sequential stages: (a) Identification, (b) Screening, (c) Eligibility (Figure 1).

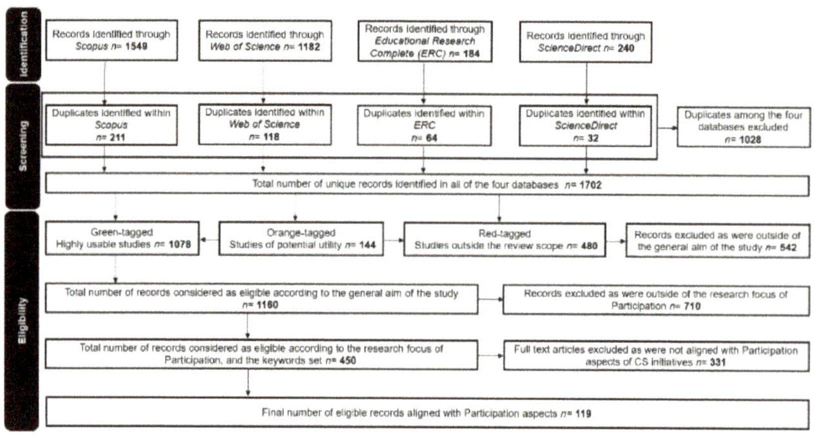

Figure 1. Flow diagram illustrating the review selection process.

This procedure resulted in 119 empirical studies, each one presenting a CS initiative; these were subjected to a content analysis, using a semi-structured checklist of CS participation-related typologies.

3. Findings

The majority of the reviewed CS initiatives were contributory and involved a single person as the data collector, mainly in the form of volunteering. Additionally, it was found that citizens were reporting multiple entries over an extended time period and were engaged only with data, contributing mainly to conducting and monitoring observations, along with collecting and submitting data, rather than being actively engaged with the whole process for a minimum duration (Figure 2).

Most of the studied CS initiatives targeted the "individual" and "private sphere" actions of EEC, which primarily took place on the "local" scale. The three main EEC outcomes reported in the reviewed CS initiatives were the development of a healthy relationship with nature, the solution of environmental problems, and the prevention of new environmental problems (Figure 3).

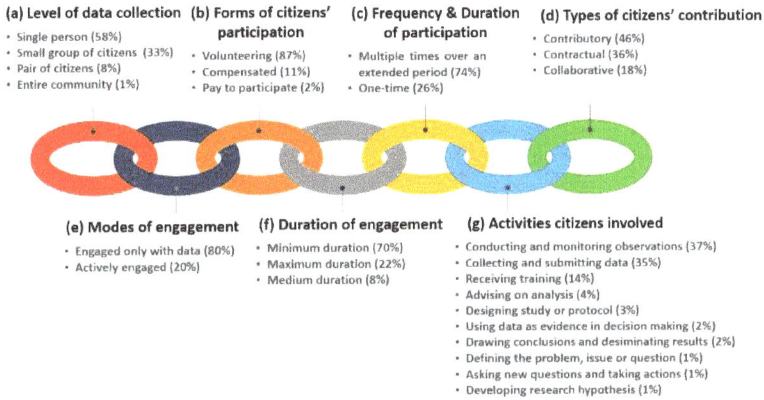

Figure 2. (**a**) Level of data collection (n = 68), (**b**) Forms of citizens' participation (n = 74), (**c**) Frequency and duration (n = 68) of citizens' participation, (**d**) Types of citizens' contributions (n = 65), € Modes of engagement (n = 57), (**f**) Duration of engagement (n = 50), and (**g**) Citizens' activities (n = 63) in CS initiatives.

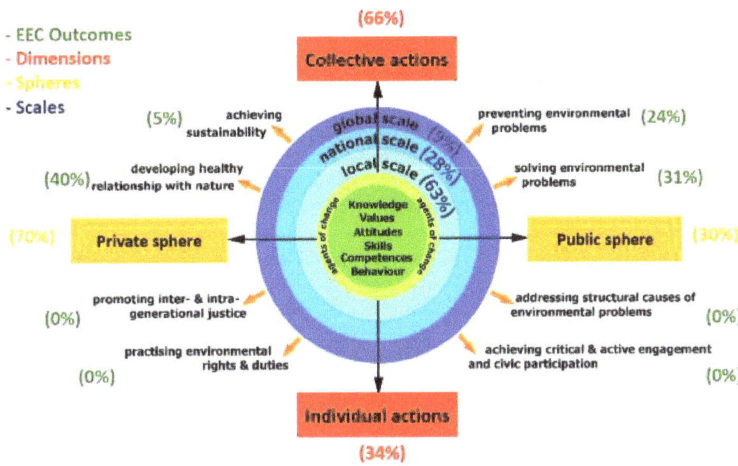

Figure 3. Dimensions, spheres, and scales as environmental citizenship actions (n = 33), as well as environmental citizenship outcomes (n = 17) promoted by the reviewed CS initiatives (Source Modified: [10]).

4. Conclusions

Our systematic literature review provides empirical substantiation on which environmental CS initiatives can support citizens' engagement with the underlying scientific processes, while also contributing to a broader, more inclusive, and active socio-political participation of citizens [11]. However, more efforts are needed to develop environmental citizenship with more focus on EEC actions situated in the collective dimension, public sphere, as well as on the national and global scale.

Author Contributions: Conceptualization, A.C.H. and D.H.; methodology, M.A.V., A.C.H., D.H., A.A. and Y.G.; software, M.A.V.; validation, A.C.H., D.H. and Y.G.; formal analysis, A.C.H.; investigation, M.A.V. and A.A.; resources, M.A.V., A.C.H., D.H., A.A. and Y.G.; data curation, M.A.V., A.A. and D.H.; writing—original draft preparation, M.A.V.; writing—review and editing, M.A.V., A.C.H., D.H. and Y.G.; visualization, A.C.H. and D.H.; supervision, A.C.H. and D.H.; project administration,

M.A.V.; funding acquisition, A.C.H. and D.H. All authors have read and agreed to the published version of the manuscript.

Funding: This research was funded by EnviroCitizen Project. The EnviroCitizen project has received funding from the European Union's Horizon 2020 research and innovation programme under grant agreement No 872557.

Institutional Review Board Statement: Not applicable.

Informed Consent Statement: Not applicable.

Data Availability Statement: No new data were created or analysed in this study. Data sharing is not applicable to this article.

Acknowledgments: This study is partly inspired by the ENEC Cost Action-European Networks for Environmental Citizenship—CA16229 supported by COST (European Cooperation in Science and Technology—Horizon 2020).

Conflicts of Interest: The authors declare no conflict of interest.

References

1. Bonney, R.; Cooper, C.; Dickinson, J.; Kelling, S.; Phillips, T.; Rosenberg, K.V.; Shirk, J. Citizen Science: A Developing Tool for Expanding Science Knowledge and Scientific Literacy. *BioScience* **2009**, *59*, 977–984. [CrossRef]
2. Fritz, S.; See, L.; Carlson, T.; Haklay, M.M.; Oliver, J.L.; Fraisl, D.; Mondardini, R.; Brocklehurst, M.; Shanley, L.A.; Schade, S.; et al. Citizen Science and the United Nations Sustainable Development Goals. *Nat. Sustain.* **2019**, *2*, 922–930. [CrossRef]
3. Irwin, A. *Citizen Science: A Study of People, Expertise, and Sustainable Development*; Routledge: London, UK, 1995; p. 6.
4. Oesterle, J.; Upadhyay, B.; Brown, J.C.; Vernon, M. Citizen Science: A Path to Democratic and Sociopolitically Conscious Science. In *Handbook of Theory and Research in Cultural Studies and Education*; Springer: Cham, Switzerland, 2020; pp. 1–30.
5. Wiggins, A.; Crowston, K. Goals and Tasks: Two Typologies of Citizen Science Projects. In Proceedings of the 2012 45th Hawaii International Conference on System Sciences, Maui, HI, USA, 4–7 January 2012; pp. 3426–3435.
6. Dickinson, J.L.; Shirk, J.; Bonter, D.; Bonney, R.; Crain, R.L.; Martin, J.; Phillips, T.; Purcell, K. The Current State of Citizen Science as a Tool for Ecological Research and Public Engagement. *Front. Ecol. Environ.* **2012**, *10*, 291–297. [CrossRef]
7. Pretty, J.; Ball, A.; Benton, T.; Guivant, J.; Lee, D.R.; Orr, D.; Pfeffer, M.J.; Ward, H. *The SAGE Handbook of Environment and Society*; SAGE Publications Ltd.: London, UK, 2021.
8. Yadav, P.; Darlington, J. Conceptual Frameworks for Building Online Citizen Science Projects. *Hum. Comput.* **2016**, *3*, 213–223. [CrossRef]
9. Jordan, R.C.; Gray, S.A.; Howe, D.V.; Brooks, W.R.; Ehrenfeld, J.G. Knowledge Gain and Behavioral Change in Citizen-Science Programs. *Conserv. Biol.* **2011**, *25*, 1148–1154. [CrossRef] [PubMed]
10. Hadjichambis, A.C.; Paraskeva-Hadjichambi, D. Education for Environmental Citizenship: The Pedagogical Approach. In *Conceptualizing Environmental Citizenship for 21st Century Education*; Hadjichambis, A.C., Reis, P., Paraskeva-Hadjichambi, D., Cincera, J., Boeve-de Pauw, J., Gericke, N., Knippels, M.-C., Eds.; Springer International Publishing: Cham, Switzerland, 2020; pp. 237–261, ISBN 978-3-030-20249-1.
11. Kenyon, E.; Christoff, A.; Wisdom, S. *Citizen Science: Expanding Ideas of Citizenship and Science*; Social Studies Research and Practice; Emerald Publishing Limited: Bingley, UK, 2020; Volume 15, pp. 83–96.

environmental sciences proceedings

MDPI

Proceeding Paper

Change the Story, Learning to Make a Difference in Climate Crisis †

Luca Baglivo and Daniela Conti *

Centre for Environmental Research, Documentation and Education CREDA, 20900 Monza, Italy; luca.baglivo@creda.it

* Correspondence: daniela.conti@creda.it

† Presented at the 2nd International Conference of International Researchers of the Education for Environmental Citizenship 2022, 10–11 March 2022. Available online: https://enec-cost.eu/ireec22/.

Abstract: We describe the results of the Italian piloting of Change the Story, a cross-European project that takes an innovative look at climate and citizenship education. The project developed resources to support teachers and pupils in creating digital narratives about the kind of world they want as a response to the climate crisis, with the aim of both improving scientific enquiry and developing citizenship competences to actively contribute to building a climate-neutral society.

Keywords: climate education; environmental citizenship; action competences; sustainability inquire learning; digital storytelling; place-based education

Citation: Baglivo, L.; Conti, D. Change the Story, Learning to Make a Difference in Climate Crisis. *Environ. Sci. Proc.* **2022**, *14*, 4. https://doi.org/10.3390/environsciproc2022014004

Academic Editors: Andreas Ch. Hadjichambis, Pedro Reis, Marie-Christine Knippels, Audronė Telesiene, Daphne Goldman, Demetra Paraskeva-Hadjichambi, Jan Cincera and Kateřina Jančaříková

Published: 1 March 2022

1. Introduction

Engaging young people to become change-makers [1] in the large challenges of the climate and sustainability crisis requires new educational approaches to empower students to make decisions and move towards social change, going beyond fostering students' environmental awareness. Education has been recognised as a critical element to deliver sustainable development [2,3] (UNESCO, 2012; UNESCO, 2014) and educative institutions need to delineate pedagogies to promote the potentialities embraced by sustainability challenges to effectively work on students' action and citizenship competences [4] allowing them to be ready to build a better world for all.

In this context, a group of six European education organizations have explored this issue within the project named Change the Story, co-funded by the Erasmus+ programme of the European Union [5], which has focused on digital storytelling as a coherent frame for students' research into and exploration of climate change to empower them to become agents of change with their friends and community.

2. Research Design and Methodology

We worked together with 10 teachers and 60 students during the 2020–2021 school year. Action research [6] was set up to outline the educational foundation for intentional environmental and civic climate action within primary education settings.

Firstly, a flow of learning grounded in the inquiry-based approach [7,8], which also includes the local community as an educational resource [9] and a creative approach of digital storytelling [10], was established to guide the co-design and the evaluation of curriculum units, together with the assessment tools used to chart the competence development progress made by students, as described in Table 1.

A framework document, gathered from the widely used framework TPACK [11], has been outlined, which lays out the principles and thinking behind the design of the Change the Story learning units.

Table 1. Dimension of the competences considered in Change the Story.

Competence	Competence Description	Indicators
Pupils can find, evaluate and use information about past, present and future climate change	Search, organize and use information for specific purposes	Search information from a variety of digital or non-digital sources Evaluate authenticity, reliability, or validity to interpret or use information Use and share information effectively and ethically
Pupils can think in a critical way to show how change in people's practices is possible	Ask, enquire, synthesize, evaluate	Questioning and analyzing evidence of climate change Extend understanding of climate change Apply scientific approach to analyze, evaluate, and take actions on climate change
Pupils can work with others to develop their digital story about the climate crisis.	Participate, collaborate data	Share responsibility Value the contributions of others Demonstrate sensitivity to other audience or contexts
Pupils can create contents about and for the planet, living organisms and people, including themselves	Apply ideas in new ways, create something of value, elaborate data	Create new narratives for the future within digital and not-digital contexts
Pupils can communicate with others	Share, exchange ideas considering impact, respect for others	Produce and share coherent and cohesive contents on climate change through digital or non-digital, oral, written, or non-verbal means.
Take responsibility on how to tackle the climate crisis	Acknowledge and accept the choices made	Demonstrate responsible citizenship through actions that contribute to sustainable communities. Evaluate the impact of decisions or actions on the dignity and well-being of individuals or communities

3. Findings

Pupils showed progression in climate literacy and the development of both sustainability and digital competencies, as shown in Figure 1. Students built an interdisciplinary approach to climate crisis and were able to make connections and interrelations between facts, data and climate data visualization. The communication phase, carried out through digital storytelling, triggered the motivation to re-elaborate findings and to create an effective call to action for schoolmates and friends. In particular, the process of creating digital stories (for some examples, see Figure 2) was found to actively enable and engage pupils in:

- Researching and understanding the basic physical science basis behind climate change;
- Searching for causes, effects and solutions;
- Having real-world experiences and observing their environment and society regarding climate change;
- Reflecting and making connections within the frames of time (past, present, future) and dimension (the self, the community, the global society).
- Expressing their ideas on climate crisis and on what they want to change in order to build a climate-neutral society.
- Competences development data were collected with self-reflection tools such as the one described in Table 2. The results are summarized in Figure 3.

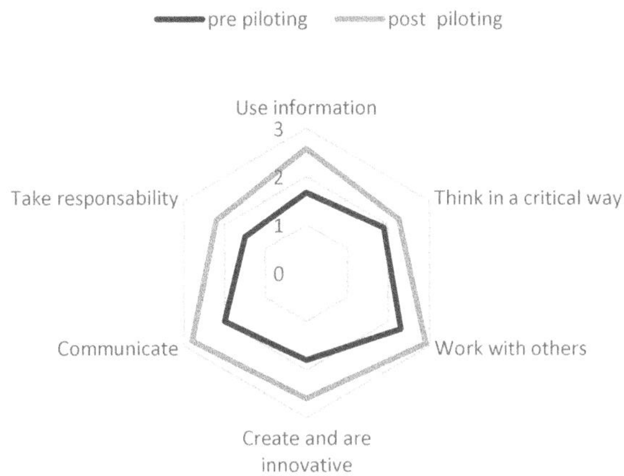

Figure 1. Results of teachers pre- and post-piloting competences assessment.

Table 2. Change the Story aimed to support pupils in developing and self-reflecting on their own digital stories about the climate crisis. Since the goal is that pupils prepare and communicate stories that are powerful to them, formative assessment tools supported content knowledge and competences development process. This example of a self-reflection tool was used throughout the process of producing a digital story about the climate crisis.

I Can	Starting	On the Way	Indipendent
Inspiring others through a digital story about the climate crisis	I still need to learn about some digital tools that help me to share my story with others.	I have learned about some digital tools. Now I need to find out how they can help me to share my story with others.	I can use digital tools to share and communicate effectively to different audiences.
Investigating the past and present	I am still learning about how to investigate and understand changes in the past, and how this influences the present climate.	I am investigating changes in the past, to find out how this influences the present climate.	I can investigate and understand changes in the past in relation to the present climate situation.
Sharing of my ideas for making changes	I still need to learn about how I can be part of making change	I am starting to learn about how to be part of making change.	I can be part of making change.
Changing things in the present for climate crisis	I still need to learn how people, including myself can make changes in the present.	I'm learning how people including myself can make a change in the present.	I'm aware ofhow people are influencing and can make changes.
Working with others	I still need to find out how I can work with others (pupils, teachers, parents, friends or other experts) to create a new story which tackles the climate crisis.	I am starting to find out how I can work with others to create a new story about what we can do about the future climate.	I can work with others to create a new story which tackles the climate crisis.

(**a**) (**b**)

Figure 2. Example of stories created by pupils. The creative components and the communication phase of the learning flow of the project were inspirational and motivating for participating students: (**a**) storyboard for the story *We are ambassadors. [we need urgently to talk to you about climate crisis]*; (**b**) *Stop to unnecessary CO_2 emission [Yes, it is still possible!]*. This video, made with Minecraft, emphasizes what adults should do to cut their carbon emissions.

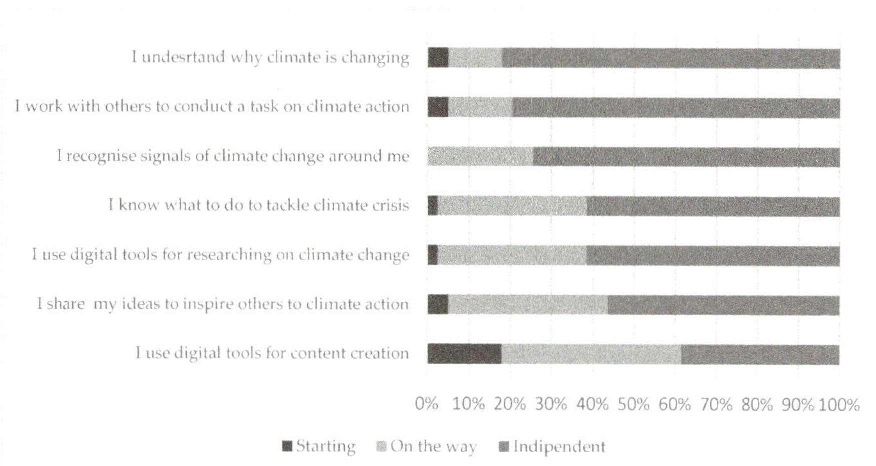

Figure 3. Results from pupils' self-assessment of content knowledge and competences development after completing the piloting.

4. Conclusions

The inquiry approach, revised to include an imaginative, creative phase for digital storytelling, was shown to be an access point for engaging in learning, enhancing scientific, digital and climate literacy, persisting in problem-solving and civic action, promoting collaboration and cooperative work between students, teachers and the community. Work on a real-world content, the climate crisis, has also received positive feedback when developed in the curricular disciplines, as it made learning more meaningful and applicable for students, thus influencing their motivation and the possibilities to learn to act for sustainability [12,13].

Author Contributions: L.B. and D.C. both contributed equally to all aspects of the paper. All authors have read and agreed to the published version of the manuscript.

Funding: This research was developed within the project Change the Story funded by the Erasmus + programme of the EC, grant number 2019-1-UK01-KA201-061432.

Institutional Review Board Statement: Not applicable.

Informed Consent Statement: Informed consent was obtained from all subjects involved in the study. No personal data was collected and all the data was anonymised.

Data Availability Statement: Data are available on request to the corresponding author.

Acknowledgments: The authors acknowledge the work of all the teachers and students who participated in the Italian piloting of the Change the Story project.

Conflicts of Interest: The authors declare no conflict of interest.

References

1. Rieckmann, M.; Mindt, L.; Gardiner, S. *Education for Sustainable Development Goals;* Learning objectives; UNESCO: Paris, France, 2017.
2. UNESCO. Shaping the education of tomorrow. In *Report on the Un Decade of Education for Sustainable Development;* UNESCO: Paris, France, 2012.
3. UNESCO. *Roadmap for Implementing the Global Action Programme on Education for Sustainable Development;* UNESCO: Paris, France, 2014.
4. Jensen, B.; Schnack, K. The action competence approach in environmental education. *Environ. Educ. Res.* **1997**, *3*, 163–178. [CrossRef]
5. Change the Story. Creating New Climate Stories across Europe. Available online: www.changethestory.eu (accessed on 16 December 2021).
6. Feldman, A.; Altrichter, H.; Posch, P.; Somekh, B. Teachers Investigate Their Work. In *An Introduction to the Methods of Action Research Across the Professions;* Routledge: London, UK, 2007.
7. Bybee, R.; Taylor, J.J.A.; Gardner, A.; Van Scotter, P.; Carlson, J.; Westbrook, A.; Landes, N. *The BSCS 5E Instructional Model: Origins and Effectiveness;* BSCS: Colorado Springs, CO, USA, 2006.
8. Minner, D.D.; Levy, A.J.; Century, J. Inquiry-based science instruction—What is it and does it matter? Results from a research synthesis years 1984 to 2002. *J. Res. Sci. Teach.* **2010**, *47*, 474–496. [CrossRef]
9. Smith, G.; Sobel, D. *Place- and Community-Based Education in Schools;* Routledge, Taylor & Francis Group: Abingdon, UK, 2010.
10. Boy, A.G. From STEM to STEAM: Toward a human-centred education, creativity & learning thinking. In Proceedings of the 31st European Conference on Cognitive Ergonomics (ECCE'13), Toulouse, France, 26–28 August 2013; Association for Computing Machinery: New York, NY, USA; Article 3, pp. 1–7. [CrossRef]
11. Mishra, P.; Koehler, M.J. Technological Pedagogical Content Knowledge: A Framework for Teacher Knowledge. *Teach. Coll. Rec.* **2006**, *108*, 1017–1054. [CrossRef]
12. Læssøe, J. Education for sustainable development, participation and socio-cultural change. *Environ. Educ. Res.* **2010**, *16*, 39–57. [CrossRef]
13. Mogensen, F.; Schnack, K. The action competence approach and the "new" discourses of education for sustainable development, competence and quality criteria. *Environ. Educ. Res.* **2010**, *16*, 59–74. [CrossRef]

environmental sciences proceedings

MDPI

Proceeding Paper

The Spatial Calibration of Environmental Citizenship: Identity Political Analysis of the Cycling Culture in a Small Provincial City †

Simo Häyrynen

Department of Geographical and Historical Studies, Joensuu Campus, University of Eastern Finland, FI-80101 Joensuu, Finland; simo.hayrynen@uef.fi

† Presented at the 2nd International Conference of International Researchers of the Education for Environmental Citizenship 2022, 10–11 March 2022. Available online: https://enec-cost.eu/ireec22/.

Abstract: This paper discusses the adoption of the urban cycling culture in a northern provincial town of Joensuu by analysing interviews of cyclists and the opinion letters from the local newspaper in the centre/periphery frameworks. It highlights the spatial change of environmental agenda and, thus, the local conditions of environmental citizenship from the identity political perspective.

Keywords: cycling culture; green urbanism; environmental citizenship; spatial sustainability transition; cultural ecosystem of traffic

Citation: Häyrynen, S. The Spatial Calibration of Environmental Citizenship: Identity Political Analysis of the Cycling Culture in a Small Provincial City. *Environ. Sci. Proc.* **2022**, *14*, 6. https://doi.org/10.3390/environsciproc2022014006

Academic Editors: Andreas Ch. Hadjichambis, Pedro Reis, Marie-Christine Knippels, Audronė Telesiene, Daphne Goldman, Demetra Paraskeva-Hadjichambi, Jan Cincera and Kateřina Jančaříková

Published: 2 March 2022

1. Introduction

Social and geographical studies on cycling often emphasize infrastructure efficiency and 'urban access', revealing important information about social aspects of cycling [1–4]. The cultural transition that occurs with the spread of urban cycling, not only as a practical mode of transportation, but also as an identity structure and social behavior, have been less researched. This paper theoretically discusses the scalar dynamics of environmental citizenship by asking how the urban cycling culture is applied and adopted in a remote provincial town. To answer this, it focuses on the local conditions of the formation of cycling identity, and the related role of active cyclists as agents of change.

The bicycle is one of the main symbols of green urbanism and low carbon ideology. Unlike material devices, the symbolic nature varies contextually [5]. Hence, the identity political approach to spatially changing cycling culture requires separation between 'sustainability' as a technical concept and a more culturally constructed symbol [6]. It enables analytical distinction between the rational and emotional aspects of environmental concerns, despite being practically interdependent.

The theoretical framework of the study is based on the culturally tense center/periphery relationship that links the subject with a larger discussion on the spatiality of green transition [7–9]. Green urbanism is based on value-based belief that the urban compact city appears as the location of development and as the solution to many problems—environmental, social, and economic [10]. Conceptually, its counterpart is 'coal capitalism' or 'petroculture' [11], which provide a platform for arguments against the too rapid spread of cycling culture. Another viewpoint, more common in peripheral areas, claims that green urbanism is not just an import of an ecologically rational way of life, but an import of a foreign cultural influence that does not work in the context of a small-town culture, and may end up hindering the real green transition [12].

2. Materials and Method

The conceptual topography will be applied to an empirical study on the city of Joensuu, a smallish Finnish provincial center in which cycling culture has recently reformed owing to a relatively young and educated population (universities). At the same time, Joensuu is a place where the urban structure, natural conditions, and political power ratios are different from those of a big city.

The research material is based on fifteen semi-stuctured interviews of cyclists and cycling experts, and SMS-messages about cycling in the feedback section of the local newspaper (Karjalainen) within five months of 2019. The study uses both conceptual and discoursive content analysis in the centre/periphery frameworks. The main themes of the analysis are (1) the rationalization and irrationalization of green lifestyle, (2) the socio-cultural dependencies of traffic, and (3) the peripheral components of the arguments.

3. Discussion

The site-specific analysis builds on the contextually changing competition of hegemony in an environmental agenda between the traditional/conventional thinking and a new formative thinking. Besides the common conflict between cyclists and(private) car drivers within a small town, the analysis shows that cyclists do not form a homogeneous group but a great variety when it comes to the motivation to use the device. A rough division takes place between the committed advocates of a green lifestyle and so-called regular cyclists. Yet, both act as agents of change in their own life-circles and in their own terms. As for all traffic users, different views on the function of cycling seems to be linked to different collective perceptions of human progress. Hence, the forms of daily mobility produce alternative epistemic spaces of cycling such as those where cycling is seen as an indication of socio-economic backwardness (cannot afford a car) or traditional practicality (bicycle only as a means of mobility). In their argumentations, all parties derive from the peripheral location of their home area and their own alleged underdog position in respect of the other traffic users.

Funding: This research was funded by the Academy of Finland, grant number 286733.

Institutional Review Board Statement: The project will follow the principles of good scientific practice and ethical guidelines of the Academy of Finland and the ethical principles of qualitative research. This means respect and confidentiality, safeguarding the anonymity of persons involved in interviews and publications, and openness to scrutiny and possible validation by fellow scientists. The research design, as presented in this plan, does not involve the collection of individual personal data that would enable individuals to be recognised or identified. The interviewees are asked for their permission to use the interviews for research purposes and to digitally record the interviews. The interview material is kept in the possession of the research project and is not passed on except as separately agreed with the interviewees.

Informed Consent Statement: Informed consent was obtained from all subjects involved in the study.

Data Availability Statement: The interview material is in the possession of the researcher. The newspaper material can be found in the electronic archive of the local newspaper. https://digilehti.karjalainen.fi/titles/karjalainen/3544/archive (accessed on 1 May 2019).

Conflicts of Interest: The authors declare no conflict of interest.

References

1. Jones, P.; Lucas, K. The social consequences of transport decision-making: Clarifying concepts, synthesising knowledge and assessing implications. *J. Transp. Geogr.* **2012**, *21*, 4–16. [CrossRef]
2. Fol, S. Social inequalities in urban access: Better ways of assessing transport improvements. In *Urban Access for the 21st Century: Finance and Governance Models for Transportation Infrastructure*; Lönnroth, M., Sclar, E., Wolmar, C., Eds.; Routledege: New York, NY, USA, 2014; pp. 46–86.
3. Camarero, L.; Oliva, J. Exploring the Social Face of Urban Mobility: Daily Mobility as Part of the Social Structure in Spain. *Int. J. Urban Reg. Res.* **2008**, *32*, 344–362. [CrossRef]
4. Šulc, I.; Morgado, S.; Đorđević, Z.; Gašparović, S.; Radović, V.; Keranova, D. Societal Issues and Environmental Citizenship. In *Conceptualizing Environmental Citizenship for 21st Century Education*; Hadjichambis, A., Reis, P., Paraskeva-Hadjicambi, D., Eds.; Springer: Cham, Switzerland, 2020; pp. 49–65. [CrossRef]

5. Bulkeley, H.; Paterson, M.; Stripple, J. *Towards a Cultural Politics of Climate Change: Devices, Desires and Dissent*; Cambridge University Press: Cambridge, UK, 2016.
6. King, D.; Ilbery, B. The environmental belief system of organic and conventional farmers: Evidence from central-southern England. *J. Rural Stud.* **2010**, *2*, 437–448. [CrossRef]
7. Coenen, L.; Truffer, B. Places and spaces of sustainability transitions: Geographical contributions to an emerging research and policy field. *Eur. Plan. Stud.* **2012**, *20*, 367–374. [CrossRef]
8. Eriksen, T.H. *Overheating: An Anthropology of Accelerated Change*; Pluto Press: London, UK, 2017.
9. Devine-Wright, P. Think Global, Act Local? The Relevance of Place Attachments and Place identities in a Climate Changed World. *Glob. Environ. Chang.* **2013**, *23*, 61–69. [CrossRef]
10. Norgaard, K.M. *Living in Denial: Climate Change, Emotions and Everyday Life*; MIT Press: Cambridge, MA, USA, 2011.
11. Wilson, S.; Szeman, I.; Carlson, A. On Petrocultures: Or, Why We Need to Understand Oil to Understand Everything Else. In *Petrocultures. Oil, Politics, Culture*; Wilson, S., Carlson, A., Szeman, I., Eds.; McGill-Queen's University Press: Montreal, QC, Canada; London, UK; Chicago, CA, USA, 2018; pp. 3–20.
12. Krastev, I.; Holmes, S. Explaining Eastern Europe: Imitation and Its Discontents. *J. Democr.* **2018**, *29*, 117–128. [CrossRef]

environmental sciences proceedings

MDPI

Proceeding Paper

Competences and Capabilities: A Relevant Resignification in Education for Environmental Citizenship †

Mónica Ofelia Garcia-Calvo [1,*], Isabel Garzón-Barragán [1] and Pedro Reis [2]

1 Facultad de Educación, Universidad Pedagógica Nacional, Bogotá 110221, Colombia; igarzon@pedagogica.edu.co
2 Instituto de Educação, Universidade de Lisboa, 1649-013 Lisboa, Portugal; preis@ie.ulisboa.pt
* Correspondence: mogarciac@upn.edu.co; Tel.: +57-310-213-96-38
† Presented at the 2nd International Conference of International Researchers of the Education for Environmental Citizenship 2022, 10–11 March 2022. Available online: https://enec-cost.eu/ireec22/.

Abstract: This paper proposes elements for the re-signification of the concept of competences from the human capabilities approach proposed by Martha Nussbaum, within the framework of the proposal of education for environmental citizenship developed by the European Network for Environmental Citizenship. For this purpose, the notions of competences and capabilities found in the book *Conceptualizing Environmental Citizenship For 21st Century Education* are analysed. From a more holistic perspective, the competences could integrate elements of reflection such as freedom of choice (with sensitivity to cultural pluralism), planning their lives (as part of environmental action planning), looking at the context of opportunities (linked to participation and action), emotional development, and the opportunity to enjoy pleasurable experiences. This document is part of the first author's doctoral research at the Universidad Pedagógica Nacional de Colombia.

Keywords: competences; capabilities; human capabilities approach; education for environmental citizenship

Citation: Garcia-Calvo, M.O.; Garzón-Barragán, I.; Reis, P. Competences and Capabilities: A Relevant Resignification in Education for Environmental Citizenship. *Environ. Sci. Proc.* **2022**, *14*, 8. https://doi.org/10.3390/environsciproc2022014008

Academic Editors: Andreas Ch. Hadjichambis, Marie-Christine Knippels, Audronė Telesiene, Daphne Goldman, Demetra Paraskeva-Hadjichambi, Jan Cincera and Kateřina Jančaříková

Published: 4 March 2022

Publisher's Note: MDPI stays neutral with regard to jurisdictional claims in published maps and institutional affiliations.

1. Introduction

Sometimes the word "capabilities" is used to define or exemplify what competences are. At other times, capabilities are understood as people's internal abilities. The truth is that a distinction is rarely made between capabilities and competences. In the education literature, greater emphasis has been placed on the development of competences than on the development of capabilities [1,2]. This research is concerned with redefining competences and capacities, given that the analysis of this relationship could integrate new and significant elements in the process of understanding and projecting the processes of training environmental citizens.

The human capabilities approach developed by Martha Nussbaum [3–5] is taken as a point of reference in that it values, from an integral perspective, the totality of opportunities available to a person to choose and to act in his or her specific political, social, and economic situation. In this perspective, the author proposes the interaction of three different types of capabilities: basic capabilities (innate faculties of each person), internal capabilities (traits and skills trained and developed in interaction with the environment), and combined capabilities (combination of internal capabilities and the social/political/economic conditions that allow them to function). This in principle calls into question referring only to capabilities as innate capacities as they are only one of the forms of expression of capabilities. Furthermore, it calls for the context to be considered in the process of defining capabilities. It is of no use to develop people's capability to make choices (internal capability) if they cannot exercise this capability in a context (combined capability). In this sense, the capabilities approach involves in a decisive way what a person is able of being and doing in a context, and thus the freedoms and opportunities to choose and to act.

In addition to capabilities, Nussbaum [3] proposes the term "functioning" to refer to the products or materialisations of capabilities. In other words, that which can be observed in people because they refer to the way they are or act in accordance with their capabilities. For example, if people have the capability to choose, the functioning would be that they put the capability to choose into action. From an educational perspective, this approach involves relevant reflections on the cultivation of freedom and the construction of a citizen of the world with the capability for understanding, sensitivity, and awareness of cultural difference; the capability to live in a close and respectful relationship with other species; the capability to contrast alternatives and have an active ethical enquiry; and the capability to place justice above political convenience [4].

2. Materials and Methods

This theoretical study analyses the notions of competences and capabilities found in the book *Conceptualizing Environmental Citizenship For 21st Century Education* [6] in the light of the notion of the human capabilities approach put forward by Martha Nussbaum to enrich the view of education for environmental citizenship. The analysis is based on the reflections on the coding process written by Packer [7], the suggestion of a pre-analysis process and reflections on inference proposed by Bardín [8], and the levels of description proposed by Martínez [9]. The pre-analysis stage includes an initial exercise of reviewing the frequencies of the concepts, using "competenc" and "capabilit" as incomplete search words.

3. Results and Discussion

In a first approach it was found that the word "competences" is mentioned 119 times while the word "capabilities" is mentioned only twice. The two times that the word "capabilities" is mentioned, it is again related to skills, especially cognitive skills that are required for action. This distances it from the definition of capabilities proposed by Nussbaum [4], since for the author capabilities are not simply abilities residing inside a person, but also include the freedoms or opportunities created by the combination of these personal faculties and the political, social, and economic environment.

Regarding the subject of competences, it was found that this term is related in the text analysed specially to processes of (a) knowledge building; (b) democratic activism; (c) values formation; (d) identity building; (e) types of education; and (f) evaluation. Each of these was taken as an emerging category for the analysis. Considering the 10 central capabilities proposed by Nussbaum [4], it could be inferred that, in this case, the term competences is in tune with capabilities such as capability for thought, capability for practical reason, capability for affiliation, capability for relationship with other species, and capability for control over one's environment.

3.1. Competence and Knowledge Building

This category refers explicitly to (a) an interdisciplinary and systemic approach; (b) co-production of new knowledge between experts and citizens; (c) include social knowledge; and (d) a basic understanding of the nature of science and acquiring basic inquiry skills. From the capabilities approach, this relationship between competences and knowledge could integrate into citizenship education, in addition to the need to know the history and social facts (through the cultivation of knowledge about other ways of life), a capacity for receptive imagination that allows us to understand the motives and options of people different from us, without seeing them as strangers who threaten us, but as beings who share with us many problems and opportunities [6]. Using the imagination to understand another person's emotions and desires and wishes or the freedom to imagine citizenship [5] would be an important contribution to building environmental citizenship.

3.2. Competence and Democratic Activism

This category refers to competences such as deep civic participation, empower people, responsible actions, participate in complex thinking, practise their environmental rights and duties, develop the willingness, combination of youth activism with citizen science, consider inter and intra-generation justice, the application of ethical principles, and planning and taking action on environmental issues they find relevant. These competences are in line with the capability approach in their focus on empowerment, social justice, deep civic engagement, and the application of ethical principles. However, they could also be redefined by tacitly including the cultivation of freedom. The capabilities approach is not only about participation and action, but also about paying detailed attention to respect for freedom of choice [4]. Non-action can also be a choice, as part of people's freedoms. The capability approach considers how the individual in his or her context can lead a life that he or she has reason to value [10].

3.3. Competence and Value Formation

This category includes universal values (social justice, wisdom, synergy with nature, equality, inner harmony, responsibility, creativity, self-respect, etc.), empathy towards the marginalised (empathic thinking), cooperation, collaboration, and communication. In this sense, it shares with the capabilities approach the proposal of justice as a moral end for the construction of a citizen of the world, and therefore deals with ingrained social injustice and inequality. Another common ground is that it is also decidedly value-pluralist in that it argues that capabilities that are of central importance to people are qualitatively differentiated from each other and cannot be reduced to a single numerical scale without being distorted [4]. This category also embraces two of Nussbaum's central capabilities related to the capacity for affiliation and the capacity to be able to live in a close and respectful relationship with animals, plants, and the natural world [4].

3.4. Competence and Identity Building

This section deals with two perspectives: (a) environmental identity; and (b) identity as agents of change. Identity is developed through the accumulation of experiences, stories, classroom materials, and ways of addressing recurring problems, knowledge, and competences connected with fostering environmental citizenship; also, it mentions that identities serve 'basic needs', including a sense of belonging, sense of competence, and autonomy. This category is related to the central capacity of control over one's environment and political power to participate effectively in the political decisions that govern our lives and to the protection of freedom of expression and association from the perspective of autonomy. Control, in turn, from the perspective of environmental citizenship, must be maintained through a permanent questioning of the forms of domination of the environment that have resulted in the present serious environmental crisis. Thus, it is important to include the possibility of planning one's own life (individual agency) in permanent connection with an environmental agency, in order to maintain a commitment to both the individual and the environment. It is also important to enable the contexts for effective participation on the level of equality and mutual recognition. It is not only about people having the internal capacity to participate, but also the possibilities in their environment to do so.

3.5. Competence and Types of Education

This category coincides with the capability approach in the need for education as a way to awaken critical thinking; it addresses some questions about how education and teachers can promote competences and proposes pedagogical alternatives such as project-based learning (PBL), inquiry-based learning (IBL), action-based and task-based learning, task-based interventions, curriculum in action, dealing with real controversial issues, place-based education, civic ecology education, ecojustice pedagogy, action competence, and socio-scientific inquiry-based learning. All of these are intended to empower educators to act as formative agents of environmental citizenship, through collaboration in planning, implementing, and evaluating of projects.

3.6. Competence and Assesment

This category addresses two ideas: a) students can also assess the efficiency of their applied education for environmental citizenship pedagogical approach; and (b) assessing and measuring the outcomes of education for environmental citizenship is also of crucial importance. However, it is not clear which evaluation approach is proposed from the perspective of competences in education for environmental citizenship.

Finally, from the capabilities approach, the integration of elements related to emotional development, play (being able to laugh, play, and enjoy activities), and physical and reproductive health could also be considered as important capabilities in the formation of citizens, and should therefore be taken into account in the framework of the competences.

4. Conclusions

The capabilities approach integrates elements that could be interesting to integrate in the process of education for environmental citizenship, such as freedom of choice (with sensitivity to cultural pluralism), planning their lives (as part of environmental action planning), looking at the context of opportunities (linked to participation and action), emotional development, and the opportunity to enjoy pleasurable experiences.

Although capabilities differ from competences in that they respond to an individual's internal demand to freely choose their own way of life and are not guided by external demand [10], competences and capabilities can enter a dynamic of permanent dialogue as long as they have common stakes towards the formation of an environmental citizenship, and question the instrumentalism driven by the economy worldwide [11].

Bearing in mind that capabilities refer to freedoms and opportunities, and competences are closely related to actions, competences could be conceived as functions or materialisations of capabilities. Conceiving first the notion of capacities in order to then think about possible competences could be a good way of recognising the plurality of contexts, the freedoms of individuals and reflecting on the opportunities that open up in contexts in order to exercise environmental citizenship in a real way. It is therefore a question of generating a dynamic of permanent dialogue between capacities and competences to plan more comprehensive training actions based on education for environmental citizenship.

Author Contributions: Conceptualization, M.O.G.-C., I.G.-B.; methodology, M.O.G.-C., I.G.-B.; software, M.O.G.-C.; validation, M.O.G.-C., I.G.-B., P.R.; formal analysis, M.O.G.-C., I.G.-B., P.R.; investigation, M.O.G.-C., I.G.-B.;resources, M.O.G.-C.; data curation, M.O.G.-C., I.G.-B., P.R.; writing—original draft preparation, M.O.G.-C., I.G.-B., P.R.; writing—review and editing, M.O.G.-C., I.G.-B., P.R.; visualization, M.O.G.-C., I.G.-B., P.R.; supervision, M.O.G.-C., I.G.-B., P.R.; project administration, M.O.G.-C., I.G.-B.; funding acquisition, M.O.G.-C. All authors have read and agreed to the published version of the manuscript.

Institutional Review Board Statement: Not applicable.

Informed Consent Statement: Not applicable.

Data Availability Statement: Not applicable.

Conflicts of Interest: The authors declare no conflict of interest.

References

1. Perrenoud, P. *Cuando la Escuela Pretende Preparar para la Vida: ¿desarrollar Competencias o enseñar otros Saberes?* Editorial GRAÓ, de IRIF, S.L.: Barcelona, Spain, 2012; p. 249.
2. UNESCO, Replantear la Educación ¿hacia un bien Común Mundial? Available online: http://www.unesco.org/new/fileadmin/MULTIMEDIA/FIELD/Santiago/pdf/replantear-educacion-ESP.pdf (accessed on 9 December 2021).
3. Nussbaum, M. ; *Crear Capacidades, Propuesta para el Desarrollo Humano*; Espasa Libros, S.L.U.: Barcelona, Spain, 2012; p. 266.
4. Nussbaum, M. Education and Democratic Citizenship: Capabilities and Quality Education. *J. Hum. Dev.* **2006**, *7*, 385–395. [CrossRef]
5. Nussbaum, M. *El Cultivo de la Humanidad: Una Defensa Clásica de la Reforma en la Educación Liberal*; Ediciones Paidós Ibérica, S.A.: Barcelona, Spain, 2005; p. 338.

6. Hadjichambis, A.C.; Reis, P.; Paraskeva-Hadjichambi, D.; Cincera, J.; Boeve-de Pauw, J.; Gericke, N.; Knippels, M.C. (Eds.) *Conceptualizing Environmental Citizenship for 21st Century Education*; SpringerOpen: Cham, Switzerland, 2018; p. 264.
7. Packer, M. *La Ciencia de la Investigación Cualitativa*, 2nd ed.; Ediciones Uniandes: Bogotá, Colombia, 2020; p. 632.
8. Bardín, L. *Análisis de Contenido*; Akal: Madrid, Spain, 1986.
9. Martínez, M. *La Investigación Cualitativa Etnográfica en Educación: Manual Teórico-Práctico*; Circulo de Lectura Alternativa Ltd.: Bogotá, Colombia, 2000; p. 88.
10. Lozano, J.F.; Boni, A.; Peris, J.; Hueso, A. Competencies in Higher Education: A Critical Analysis from the Capabilities Approach. *J. Philos. Educ.* **2012**, *46*, 132–146. [CrossRef]
11. Sterling, S. Assuming the Future: Repurposing Education in a Volatile Age. In *Post-Sustainability and Environmental Education*; Jickling, B., Sterling, S., Eds.; Palgrave Macmillan: Cham, Switzerland, 2017; pp. 31–45. [CrossRef]

environmental sciences proceedings

MDPI

Proceeding Paper

Developing an Evidence-Based Educational Course for Environmental Citizenship †

Mykolas Simas Poškus

Institute of Psychology, Mykolas Romeris University, LT-08303 Vilnius, Lithuania; mykolas_poskus@mruni.eu

† Presented at the 2nd International Conference of International Researchers of the Education for Environmental Citizenship 2022, 10–11 March 2022. Available online: https://enec-cost.eu/ireec22/.

Abstract: The current generation of high school students is faced with a vast torrent of information regarding current events and issues, and environmental issues tend to be among the more salient topics. The quality and veracity of the contents of the information often times is dubious, and this illustrates the need to establish a reliable source of factual and politically neutral information regarding environmental issues; this can be conveniently accomplished by introducing education for environmental citizenship into the high school curriculum in a more systemic manner.

Keywords: environmental education; longitudinal studies; intervention; evidence-based education; action-based knowledge

Citation: Poškus, M.S. Developing an Evidence-Based Educational Course for Environmental Citizenship. *Environ. Sci. Proc.* **2022**, *14*, 10. https://doi.org/10.3390/environsciproc2022014010

Academic Editors: Andreas Ch. Hadjichambis, Pedro Reis, Marie-Christine Knippels, Audronė Telesiene, Daphne Goldman, Demetra Paraskeva-Hadjichambi, Jan Cincera and Kateřina Jančaříková

Published: 7 March 2022

Introduction

Currently, the environmentalist movement is gaining increasingly more public support and attention, and for good reason. In order to tackle current environmental issues, we need society to participate actively in mitigating the harm done to the environment and to participate as citizens in working toward ways of tackling environmental issues. The activist movements associated with environmental issues, however, are often politicized in ways that are more than just a fight for the environment, which may lead some people to have aversive reactions to environmentalism altogether. This type of situation benefits no one.

Additionally, many environmental activists may have opinions regarding various environmental issues that are not factual but are based in personal opinion and feelings (e.g., blaming whole political or economic systems, regarding nuclear energy as a threat to the environment, overestimating the capacity to adopt various solutions in a short period of time, and so on). The current situation in Europe, where countries with overly ambitious environmental goals were forced to go back to coal power is a prime example of good intentions backfiring because of decisions made without rational foresight. Thus, while environmental activism is in essence a positive force for change and is necessary, it needs to stem from a strong foundation of factual knowledge.

To address the aforementioned issue, the ENVICI (http://envici.mruni.eu) project was conceived. The aim of the project is to foster the development of environmental citizenship in school-aged children through evidence-based education that focuses on providing state of the art factual knowledge regarding environmental issues to young people, both in terms of basic understanding of ecological systems and of acquiring concrete action-based knowledge regarding everyday activities that are environmentally friendly.

In the context of the project, educational materials (an exercise book suitable for school-aged children) will be developed in order to provide an engaging and effective way of transmitting factual knowledge to young people who then will have the necessary understanding to engage in societal actions as environmental citizens. This, hopefully, will lead to a more evidence-driven environmentalism that focuses on issues from a reasonable and practically possible point of view, with tangible and realistic proposals.

The effectiveness of the educational materials in shaping environmental citizens will be tested longitudinally through an intervention study. We take the model of environmental citizenship proposed by Hawthorne and Alabaster [1] as a basis for the change model we are going to use in assessing whether students' environmental citizenship has been affected by the educational materials we are developing (Figure 1). We are particularly interested in the part of the model that proposes that both abstract and concrete knowledge, as well as education and need for learning about environmentalism contributes to environmental citizenship. As identified by a survey of various stakeholders, there is a strong need for education for environmental citizenship in Lithuania [2] and across Europe [3]. Thus, high-quality, politically neutral, and factual educational materials that contribute to educating an environmental citizen seem to be needed if we are to address environmental issues in a sustainable and manageable way through the informed social participation of citizens.

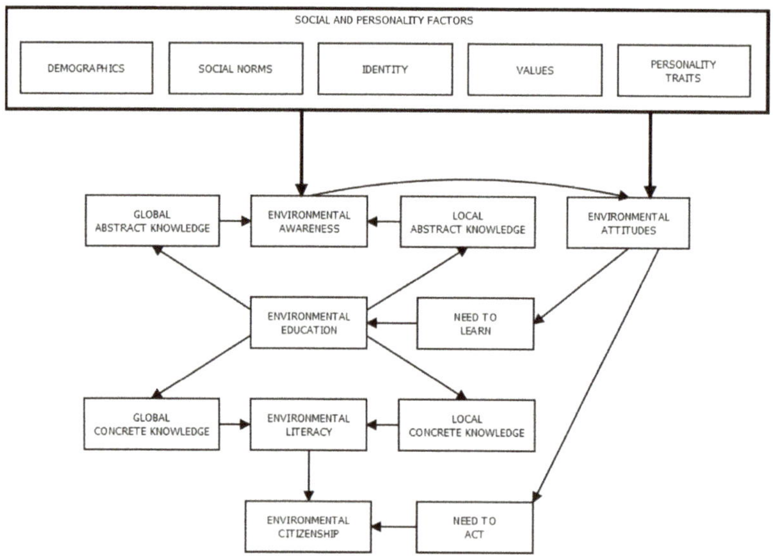

Figure 1. Theoretical model that will be used in the project.

The instruments used to assess the components of the model will be largely based on previous research on environmental citizenship [4], while some measures will be constructed by the project team as there is no clear consensus on how to assess some of the components of the model.

The intervention will be piloted on a small scale with the intention to develop the final educational materials and to assess their quality and effectiveness through a small-scale longitudinal study. After the materials are finalized, the exercise book will be finalized and tested in a large-scale longitudinal study involving several schools from Lithuania. If the data support the effectiveness of the book, it will then be made publicly available for schools to use free of charge. Additionally, translations into other languages will be considered in order to broaden the impact of the project.

Funding: This research was funded by a grant (No. S-MIP-21-60) from the Research Council of Lithuania.

Institutional Review Board Statement: Not applicable.

Informed Consent Statement: Not applicable.

Data Availability Statement: Not applicable.

Conflicts of Interest: The author declares no conflict of interest.

References

1. Hawthorne, M.; Alabaster, T. Citizen 2000: Development of a model of environmental citizenship. *Glob. Environ. Change* **1999**, *9*, 25–43. [CrossRef]
2. Poškus, M.S.; Balundė, A.; Jovarauskaitė, L. 13. SWOT Analysis of Environmental Citizenship Education in Lithuania. In *European SWOT Analysis on Education for Environmental Citizenship*; Hadjichambis, A.C., Reis, P., Paraskeva-Hadjichambi, D., Eds.; Intitute of Education—University of Lisbon, Cyprus Centre for Environmental Research and Education & European Network for Environmental Citizenship—ENEC Cost Action: Lisbon, Portugal, 2019; ISBN 978-9963-9275-6-2.
3. Hadjichambis, A.C.; Reis, P. Introduction to the Conceptualisation of Environmental Citizenship for Twenty-First-Century Education. In *Conceptualizing Environmental Citizenship for 21st Century Education. Environmental Discourses in Science Education*; Hadjichambis, A.C., Reis, P., Paraskeva-Hadjichambi, D., Činčera, J., Pauw, J.B., Gericke, N., Knippels, M.-C., Eds.; Springer: Cham, Switzerland, 2020; Volume 4, pp. 1–14.
4. Hadjichambis, A.C.; Paraskeva-Hadjichambi, D. Environmental Citizenship Questionnaire (ECQ): The Development and Validation of an Evaluation Instrument for Secondary School Students. *Sustainability* **2020**, *12*, 821. [CrossRef]

environmental sciences proceedings

MDPI

Proceeding Paper

Students' Action Competence for Sustainability and the Effectiveness of Sustainability Education †

Daniel Olsson [1,*], Niklas Gericke [1] and Jelle Boeve-de Pauw [2]

1 Institution of Environmental and Life Sciences, Karlstad University, 65188 Karlstad, Sweden; niklas.gericke@kau.se
2 Freudenthal Institute, Faculty of Science, Utrecht University, 3584 CC Utrecht, The Netherlands; j.n.a.boeve-depauw@uu.nl
* Correspondence: daniel.olsson@kau.se

† Presented at the 2nd International Conference of International Researchers of the Education for Environmental Citizenship 2022, 10–11 March 2022. Available online: https://enec-cost.eu/ireec22/.

Abstract: Scholarly attention has recently been increasingly focused on the concept of action competence for sustainability and its importance to promote environmental citizenship. Still, knowledge about the effects of sustainability education (SE) as an approach to teaching to foster students' environmental citizenship in terms of action competence for sustainability, where SE could be defined by holism (the approach to the sustainability content) and pluralism (the approach to teaching). The aim of this study is therefore to contribute new knowledge of effects of SE on young people's self-perceived action competence for sustainability (SPACS), through a longitudinal design. Our results show that SE as a teaching approach is effective in fostering environmental citizenship in terms of the important aspect of action competence for sustainability.

Keywords: longitudinal; action competence; environmental citizenship; holism and pluralism

Citation: Olsson, D.; Gericke, N.; Pauw, J.B.-d. Students' Action Competence for Sustainability and the Effectiveness of Sustainability Education. *Environ. Sci. Proc.* **2022**, *14*, 11. https://doi.org/10.3390/environsciproc2022014011

Academic Editors: Andreas Ch. Hadjichambis, Pedro Reis, Marie-Christine Knippels, Audronė Telesiene, Daphne Goldman, Demetra Paraskeva-Hadjichambi, Jan Cincera and Kateřina Jančaříková

Published: 9 March 2022

1. Introduction

Scholarly attention has recently been focused on the concept of action competence for sustainability and its importance to promote environmental citizenship [1]. Still, knowledge about the effects of sustainability education (SE) as an approach to foster students' environmental citizenship in terms of action competence for sustainability, where SE could be defined by holism (the approach to the sustainability content) and pluralism (the approach to teaching) [2]. The aim of this study is, therefore, to contribute new knowledge of the effects of SE on young people's self-perceived action competence for sustainability (SPACS). Through a longitudinal design, we followed students in a school where the teachers participated in a school development project aiming to implement SE. Two questions are posed:

- What is the development of students' SPACS and their experience of the SE teaching?
- What is the effect of SE on SPACS?

2. Methods

Student questionnaire data were collected in three waves (September 2017–June 2019) in accordance with a longitudinal design. In total, 760 Swedish upper secondary students participated in the waves, starting in grade 10 and 11, through grades 11 and 12 in the second and the final wave. Two scales were used, the SPACS-scale ($\alpha = 0.90$) [3] and the ESD, holism, and pluralism scale ($\alpha = 0.80$) [2]. Our data were analyzed using latent growth modeling and structural equation modeling. Measurement invariance between the waves was established using multiple fit indices.

3. Results

The longitudinal results reveal a significant increase for students' experience of holism, but not for their experience of pluralism. This result show that the students have experienced something in the sustainability teaching, especially in relation to the approach to the content, but not in relation to the pluralistic approach in teaching to the same degree. The results also reveal a significant increase over time regarding students' SPACS. Finally, our SEM analyses show that experiencing the SE dimensions of holism and pluralism at the student level will have a positive effect on their action competence for sustainability (self-perceived) (β between 0.14 and 0.30).

4. Conclusions and Recommendations

Our results show that SE as a teaching approach is effective in fostering environmental citizenship in terms of the important aspect of action competence for sustainability. Moreover, the results shed light on the fact that a SE school development project to some degree can support the development of students' experiences of SE teaching at their school, which then positively influence their self-perceived action competence for sustainability.

Author Contributions: This conference paper is the result of a collaborative work where all the authors significantly contributed to its conceptualization, methodology, validation, formal analysis, data curation, and the writing. All authors have read and agreed to the published version of the manuscript.

Funding: This work was supported by the Swedish Institute for Educational Research under Grant [number 2017-00065].

Institutional Review Board Statement: The study follows the ethical guidelines in Sweden and the specific guidelines provided by Karlstad University.

Informed Consent Statement: The participants were informed about the purpose of the project and they all provided active informed consent for participating.

Data Availability Statement: Data supporting the findings of this study are available from the corresponding author upon reasonable request.

Acknowledgments: We are grateful to Teresa Berglund at Karlstad University for sharing her expertise and for her important contribution during the development of this article. This research was supported by ROSE (Research On Subject-specific Education), Karlstad University.

Conflicts of Interest: The authors declare no conflict of interest. The funders had no role in the design of the study, in the collection, analyses, or interpretation of data, in the writing of the manuscript, or in the decision to publish the results.

References

1. Hadjichambis, A.C.; Reis, P.; Paraskeva-Hadjichambi, D.; Činčera, J.; Boeve-de Pauw, J.; Gericke, N.; Knippels, M.C. *Conceptualizing Environmental Citizenship for 21st Century Education*; Springer: Berlin/Heidelberg, Germany, 2020.
2. Boeve-de Pauw, J.; Gericke, N.; Olsson, D.; Berglund, T. The Effectiveness of Education for Sustainable Development. *Sustainability* **2015**, *7*, 15693–15717. [CrossRef]
3. Olsson, D.; Gericke, N.; Sass, W.; Pauw, J.B.-D. Self-perceived action competence for sustainability: The theoretical grounding and empirical validation of a novel research instrument. *Environ. Educ. Res.* **2020**, *26*, 742–760. [CrossRef]

environmental sciences proceedings

MDPI

Proceeding Paper

The Role of Environmental Journalism and Documentaries as a Means of Informal Education for Environmental Citizenship †

María Purificación Subires-Mancera [1,*] and José Jesús Delgado-Peña [2,*]

[1] Department of Journalism, Faculty of Communication Sciences, Campus de Teatinos, University of Málaga, 29071 Málaga, Spain
[2] Department for Didactics of Mathematics, Social Sciences and Experimental Sciences, Faculty of Education Sciences, Campus de Teatinos, University of Málaga, 29071 Málaga, Spain
* Correspondence: purificacion@uma.es (M.P.S.-M.); jdelgado@uma.es (J.J.D.-P.)
† Presented at the 2nd International Conference of International Researchers of the Education for Environmental Citizenship 2022, 10–11 March 2022. Available online: https://enec-cost.eu/ireec22/.

Abstract: The documentary is one of the genres most commonly used in Environmental Journalism to inform, educate and raise awareness among citizens about the conservation and defense of the environment. The objective of this paper is to analyze the role of documentaries as tools for Informal Environmental Education, by means of case studies, in order to examine how they can contribute to Environmental Citizenship. The main conclusions that can be highlighted are that the in-depth treatment of the issues, showing the images that reflect the attacks against nature and using expert scientific sources make the public know the problems, reflect on them and develop a critical awareness.

Keywords: documentary; informal education; environmental citizenship; mass media; environmental journalism

Citation: Subires-Mancera, M.P.; Delgado-Peña, J.J. The Role of Environmental Journalism and Documentaries as a Means of Informal Education for Environmental Citizenship. *Environ. Sci. Proc.* **2022**, *14*, 12. https://10.3390/environsciproc2022014012

Academic Editors: Andreas Ch. Hadjichambis, Pedro Reis, Marie-Christine Knippels, Audronė Telesiene, Daphne Goldman, Demetra Paraskeva-Hadjichambi, Jan Cincera and Kateřina Jančaříková

Published: 9 March 2022

1. Introduction

The media fulfill the triple function of educating, informing and entertaining, acting as agents of socialization and as channels for the Informal Education of citizens. In the case of Environmental Education, the role of Environmental Journalism should be highlighted, because it is an area of specialization that has a marked didactic function, since it not only informs and denounces damages and attacks against nature, but also seeks to educate and raise awareness among citizens, so that they become committed to the conservation and defense of the environment (Esteve and Fernández del Moral, 1999) [1]. One of the genres most commonly used in Environmental Journalism to the compliance of these functions is the documentary. It is a genre that has a double informative and didactic aspect, which addresses the issues in depth, from multiple viewpoints and in a critical manner, with the aim of making the public reflect on them (Rabiger, 1989) [2]. These characteristics make the documentary an appropriate informal educational resource for the environmental awareness of citizens.

2. Objective, Materials and Methods

The objective of this paper is to analyze the role of documentaries as tools for Informal Environmental Education, by means of case studies, in order to examine how they can contribute to Environmental Citizenship. The main elements under study are the subjects addressed (what is expounded), the sources utilized (who talks about the topic) and the resources used for environmental education (how it is explained). The documentaries studied are 'Buy, throw away, buy. The secret history of planned obsolescence' [3] (Cosima Dannoritzer, 2010), 'Sonic Sea' [4] (Michelle Dougherty and Daniel Hinerfeld, 2016), 'A Plastic Ocean' [5] (Craig Leeson, 2016), 'Climate Change: The Facts' [6] (Serena Davies, 2019) and the web documentary 'No Ecological Footprint' [7] (Lab RTVE, 2018).

3. Results

Each documentary focuses on an aspect of interest for Environmental Education: planned obsolescence of producing, reusing and recycling; ocean noise pollution and how it seriously harms whales and other marine species; plastic pollution of the oceans; climate change; and the ecological footprint and ways to reduce our footprint.

4. Conclusions

The main conclusion that can be highlighted is the role of documentaries in the Environmental Education of citizens, since the in-depth treatment of the issues, showing the images that reflect the attacks against nature and using expert scientific sources make the public know the problems, reflect on them and develop a critical awareness.

Institutional Review Board Statement: Not applicable.

Informed Consent Statement: Not applicable.

Conflicts of Interest: The authors declare no conflict of interest.

References

1. Esteve, F.; Fernández del Moral, J. *Áreas de Especialización Periodística*; Fragua: Madrid, Spain, 1999.
2. Rabiger, M. *Dirección de Documentales*; Instituto Oficial de Radio y Televisión RTVE: Madrid, Spain, 1989.
3. Comprar, Tirar, Comprar. La Historia Secreta de la Obsolescencia Programada. Available online: https://www.rtve.es/television/documentales/comprar-tirar-comprar/ (accessed on 30 October 2021).
4. Sonic Sea. Available online: https://sonicsea.org/ (accessed on 30 October 2021).
5. A Plastic Ocean. Available online: https://plasticoceans.org/ (accessed on 30 October 2021).
6. El Cambio Climático: Los Hechos. Available online: https://www.rtve.es/play/videos/somos-documentales/cambio-climatico-hechos/5738637/ (accessed on 30 October 2021).
7. No Ecological Footprint. Available online: https://lab.rtve.es/huella-ecologica/en/ (accessed on 30 October 2021).

environmental sciences proceedings

MDPI

Proceeding Paper

Key Pedagogical Features and a Common Approach to Evaluate Education for Environmental Citizenship: An International Perspective†

Marta Romero Ariza [1,*], Jelle Boeve-de Pauw [2,3], Daniel Olsson [4], Peter Van Petegem [2], Gema Parra [5] and Niklas Gericke [4]

1 Department of Didactics of Sciences, University of Jaén, 23071 Jaen, Spain
2 Research Unit Edubron, Department of Training and Education Sciences, Faculty of Social Sciences, University of Antwerp, 2000 Antwerp, Belgium; jelle.boeve-depauw@uantwerpen.be (J.B.-d.P.); peter.vanpetegem@uantwerpen.be (P.V.P.)
3 Freudenthal Institute, Faculty of Science, Utrecht University, 3584 CC Utrecht, The Netherlands
4 Department of Environmental and Life Science, Karlstad University, 65188 Karlstad, Sweden; daniel.olsson@kau.se (D.O.); niklas.gericke@kau.se (N.G.)
5 Animal Biology, Plant Biology and Ecology Department, University of Jaén, 23071 Jaen, Spain; gparra@ujaen.es
* Correspondence: mromero@ujaen.es
† Presented at the 2nd International Conference of International Researchers of the Education for Environmental Citizenship 2022, 10–11 March 2022. Available online: https://enec-cost.eu/ireec22/.

Citation: Ariza, M.R.; Pauw, J.B.-d.; Olsson, D.; Van Petegem, P.; Parra, G.; Gericke, N. Key Pedagogical Features and a Common Approach to Evaluate Education for Environmental Citizenship: An International Perspective. *Environ. Sci. Proc.* **2022**, *14*, 13. https://doi.org/10.3390/environsciproc2022014013

Academic Editors: Andreas Ch. Hadjichambis, Pedro Reis, Marie-Christine Knippels, Audronė Telesiene, Daphne Goldman, Demetra Paraskeva-Hadjichambi, Jan Cincera and Kateřina Jančaříková

Published: 10 March 2022

Publisher's Note: MDPI stays neutral with regard to jurisdictional claims in published maps and institutional affiliations.

Abstract: This paper presents various educational interventions aimed at promoting environmental citizenship, which were developed in three different European countries (Sweden, Belgium and Spain). The interventions differ in context, target group and educational setting (formal or non-formal) and were evaluated in terms of their impact on participants' knowledge, attitudes and behaviours. The results show significant differences between pre and post scores, with a positive impact on the behavioural dimension in all of the reported interventions. Finally, the interventions are discussed on the basis of key common pedagogical features aligned with the specialised literature.

Keywords: sustainability consciousness questionnaire (SCQ); education for environmental citizenship; evaluation of interventions; formal education; non-formal education

1. Rational and Objectives

Several authors have suggested approaching environmental problems through citizenship [1–3], and important efforts have been made to conceptualise the idea of environmental citizenship [4].

Education is considered a main tool for capacity building and for providing citizens with fundamental knowledge and meaningful opportunities to exercise action competences to actively contribute to the generation of sustainable solutions to current and future problems.

Different pedagogical approaches with high potential to promote environmental citizenship have been identified. Činčera et al. [5] maintain that these interventions that have been proven to have a significant impact on people's beliefs, attitudes and behaviours exhibit common features: they engage individuals in the collaborative construction of sustainable solutions to local problems, fostering a sense of ownership and empowerment towards environmental issues.

The model developed by Paraskeva-Hadjichambi is aligned with the key pedagogical features previously mentioned [6], including processes such as inquiry, planning, acting, evaluating, reflecting and use dissemination and networking to enhance the effect of those interventions. Nevertheless, there is a need to better understand how this educational model might be implemented in different contexts and situations, as well as to evaluate the impact of these interventions on environmental citizenship. The Sustainability Consciousness Questionnaire (SCQ) has been proven to be a powerful instrument for this purpose [7].

Addressing all these concerns, this work sets the following objectives:

1. To discuss educational interventions taking place in three countries differing in context, target group and educational setting (formal or non-formal).
2. To evaluate these interventions according to their impact on participants' knowledge, attitudes and behaviours.
3. To identify key pedagogical features common to the different interventions and to discuss to what extent these features are aligned with the specialised literature.

2. Research Design and Methodology

A single-group, pre and post test research design was used to measure the effect of various educational interventions on participants' beliefs, attitudes and self-reported behaviours, using an instrument previously validated in the specialised literature [8].

3. Findings and Conclusions

The results show significant differences between pre and post scores, with a positive impact on the behavioural dimension in all the reported interventions. Even though the three cases presented addressed different target groups, took place in very different contexts and varied in length, we can find common pedagogical features: the three of them promoted active and situated learning and were contextualised in real-life problems, offering meaningful opportunities for action-taking and reflection.

Author Contributions: This paper is the result of a truly collaborative work where all the authors (M.R.A., J.B.-d.P., D.O., P.V.P., G.P. and N.G.) significantly contributed to its conceptualisation, methodology, validation, formal analysis, data curation, and the writing, editing and reviewing process when preparing the present manuscript. All authors have read and agreed to the published version of the manuscript.

Funding: The Belgian case was part of the VALIES project and was supported by the Flanders Research Foundation (Fonds voor Wetenschappelijk Onderzoek, FWO) under Grant number S010317N within the Strategic Basic Research program. The Swedish case was funded by The Swedish Institute for Educational Research (grant number: 2017-00065). We received a fee to cover the open-access costs for publication in *Sustainability* from the ENEC COST action CA16229.

Institutional Review Board Statement: In the Belgian case, data were collected under the positive advice given by the ethical committee for social sciences and humanities to the VALIES project (SHW_18_25). In the Spanish case, data collection follows the guidelines provided by the ethical committee for social sciences and humanities of the University of Jaén. In the Swedish case, the study follows the ethical guidelines provided by Karlstad University.

Informed Consent Statement: All Belgian participants provided active informed consent; for minors, consent was received from a parent or legal guardian combined with the children providing informed assent. The Spanish participants were informed about the purpose of data collection, accepted to participate and provided patient consent. The Swedish participants were informed about the purpose of the project, and they all provided active informed consent for participating.

Data Availability Statement: Data supporting the findings of this study are available from the corresponding author upon reasonable request.

Acknowledgments: This article is based upon work from COST Action European Network for Environmental Citizenship—ENEC CA16229, supported by COST (European Cooperation in Science and Technology) https://www.cost.eu/actions/CA16229/ (accessed on 2 March 2022).

Conflicts of Interest: The authors declare no conflict of interest. The funders had no role in the design of the study, in the collection, analyses, or interpretation of data, in the writing of the manuscript, or in the decision to publish the results.

References

1. Beck, U. Climate for Change, or how to Create a Green Modernity? *Theory Cult. Soc.* **2010**, *27*, 254–266. [CrossRef]
2. Bell, D.R. Liberal Environmental Citizenship. *Environ. Politics* **2005**, *14*, 179–194. [CrossRef]
3. Cao, B. *Environment and Citizenship*; Routledge: New York, NY, USA, 2015.
4. ENEC. European Network for Environmental Citizenship. In Defining "Education for Environmental Citizenship"; EU: 2018. Available online: https://enec-cost.eu/our-approach/education-for-environmental-citizenship/ (accessed on 9 March 2022).
5. Činčera, J.; Romero-Ariza, M.; Zabic, M.; Kalaitzidaki, M.; del Consuelo Díez Bedmar, M. Environmental Citizenship in Primary Formal Education. In *Conceptualizing Environmental Citizenship for 21st Century Education*; Springer: Berlin/Heidelberg, Germany, 2020; pp. 163–177.
6. Hadjichambis, A.C.; Paraskeva-Hadjichambi, D. Education for Environmental Citizenship: The Pedagogical Approach. *Environ. Discourses Sci. Educ.* **2020**, *4*, 237–261.
7. Ariza, M.R.; Boeve-de Pauw, J.; Olsson, D.; Van Petegem, P.; Parra, G.; Gericke, N. Promoting Environmental Citizenship in Education: The Potential of the Sustainability Consciousness Questionnaire to Measure Impact of Interventions. *Sustainability* **2021**, *13*, 11420. [CrossRef]
8. Gericke, N.; Boeve-de Pauw, J.; Berglund, T.; Olsson, D. The Sustainability Consciousness Questionnaire: The Theoretical Development and Empirical Validation of an Evaluation Instrument for Stakeholders Working with Sustainable Development. *Sustain. Dev.* **2019**, *27*, 35–49. [CrossRef]

environmental sciences proceedings

MDPI

Proceeding Paper

Knowledge Use and Environmental Education in Hungarian School Gardens †

Imre Kovách [1] and Boldizsár Megyesi [2,*]

1 Institute of Sociology, Centre for Social Sciences, University of Debrecen, 4032 Debrecen, Hungary; kovach.imre@tk.mta.hu
2 Institute of Sociology, Centre for Social Sciences, Hungarian Academy of Sciences, 1097 Budapest, Hungary
* Correspondence: megyesi.boldizsar@tk.mta.hu

† Presented at the 2nd International Conference of International Researchers of the Education for Environmental Citizenship 2022, 10–11 March 2022. Available online: https://enec-cost.eu/ireec22/.

Abstract: Environmental education, as defined by Hadjichambis et al., is effective when it combines knowledge types as well as organizational and management forms. Formal, informal and non-formal education are mediators of other types of knowledge, and participants have different perceptual interests and intentions and motivations. This paper focuses on an example of a non-formal environmental education form. It presents and analyses the types of knowledge and the motivations for their use in environmental education in Hungary in the example of school gardens.

Keywords: school garden; knowledge forms; power and knowledge; environmental education; motivations

Citation: Kovách, I.; Megyesi, B. Knowledge Use and Environmental Education in Hungarian School Gardens. *Environ. Sci. Proc.* **2022**, *14*, 14. https://doi.org/10.3390/environsciproc2022014014

Academic Editors: Pedro Reis, Marie-Christine Knippels, Audronė Telesiene, Daphne Goldman, Demetra Paraskeva-Hadjichambi, Jan Cincera and Kateřina Jančaříková

Published: 10 March 2022

1. Introduction

Smederevac et al. [1] pointed out that environmental education, as defined by Hadjichambis et al. [2], is effective when it combines knowledge types as well as organizational and management forms, whereas formal, informal and non-formal education are mediators of other types of knowledge and participants have different perceptual interests and intentions and motivations.

This paper focuses on an example of a non-formal environmental education form. It presents and analyses the types of knowledge and the motivations for their use in environmental education in Hungary in the example of school gardens.

2. Methods and Materials

The paper is based on document analysis (policy documents, planning documents), of available data and semi-structured interviews, conducted with teachers responsible for school gardens, civic organizers of the School Garden movement and an additional interview with a representative of the public administration. The interviews were transcribed. We analysed the interviews using a semi-open-coded method to explore power relations and knowledge forms which influence the development of school gardens.

3. Results

The first part of this paper is about the theories of the types of knowledge [3] and the relationships between power and knowledge use [4]. The number of school gardens has grown steadily in recent decades, previously created and managed by enthusiastic teachers and their pedagogical allies following the ethos of scientific knowledge [5], with emphasis on the importance and pedagogical usefulness of traditional, local, tacit knowledge [6,7]. The School Garden movement was later founded, which is also supported by the President of the Republic's Blue Planet Foundation, the Ministry of Agriculture, and the Chamber of Agriculture, along with other organizations such as churches. Sponsors also provide financial resources, and this has been accompanied by a gradual advance in managerial knowledge of the project class [8].

4. Discussion

The second part of this paper presents the case studies and the processes that are important for educating for environmental citizenship: through examples, children learn about environmental responsibility and healthy food, and show how the knowledge gained in SE classes can be turned into a real experience.

5. Conclusions

The third part of this paper concludes the analyses on forms of knowledge used in school garden practices, showing the direction in which this version of non-formal education has changed with the involvement of state, ministerial and foundation supporters, and the power and interest aspirations lined up alongside the original motivations.

Boldizsár Megyesi is supported by the Bolyai János Post-doctoral stipendium and the ÚNKP 2020-XX-5.

Author Contributions: Conceptualization, I.K. and B.M.; methodology, I.K. and B.M.; validation, I.K. and B.M.; formal analysis, I.K. and B.M.; investigation, I.K. and B.M.; resources, I.K. and B.M.; writing—original draft preparation, I.K. and B.M.; writing—review and editing, I.K. and B.M. All authors have read and agreed to the published version of the manuscript.

Institutional Review Board Statement: Not applicable.

Informed Consent Statement: Not applicable.

Data Availability Statement: Not applicable.

Conflicts of Interest: The authors declare no conflict of interest.

References

1. Smederevac-Lalic, M.; Finger, D.; Kovách, I.; Lenhardt, M.; Petrovic, J.; Djikanovic, V.; Conti, D.; Boeve-de Pauw, J. Knowledge and Environmental Citizenship. In *Conceptualizing Environmental Citizenship for 21st Century Education*; Hadjichambis, A.C., Reis, P., Paraskeva-Hadjichambi, D., Činčera, J., Boeve-de Pauw, J., Gericke, N., Knippels, M.-C., Eds.; Springer International Publishing: Cham, Switzerland, 2020; pp. 69–82, ISBN 978-3-030-20249-1.
2. Lenhardt, M.; Smederevac-Lalic, M.; Radovic, V. *European SWOT Analysis on Education for Environmental Citizenship*; Hadjichambis, A.C., Reis, P., Paraskeva-Hadjichambi, D., Eds.; Short Country Report SERBIA; Intitute of Education, University of Lisbon: Lisbon, Portugal, 2019; pp. 246–260.
3. Bruckmeier, K.; Tovey, H. Knowledge in Sustainable Rural Development: From Forms of Knowledge to Knowledge Processes. *Sociol. Rural.* **2008**, *48*, 313–329. [CrossRef]
4. Csurgó, B.; Kovách, I.; Kučerová, E. Knowledge, Power and Sustainability in Contemporary Rural Europe. *Sociol. Rural.* **2008**, *48*, 292–312. [CrossRef]
5. Rahman, A. *Development of an Integrated Traditional and Scientific Knowledge Base: A Mechanism for Accessing, Benefit-Sharing and Documenting Traditional Knowledge for Sustainable*; UNCTAD: Geneva, Switzerland, 2000.
6. Folke, C. Traditional Knowledge in Social–Ecological Systems. *Ecol. Soc.* **2004**, *9*, 7. [CrossRef]
7. Janssen, M.A.; Anderies, J.M.; Ostrom, E. Robustness of Social-Ecological Systems to Spatial and Temporal Variability. *Soc. Nat. Resour.* **2007**, *20*, 307–322. [CrossRef]
8. Kovách, I.; Kučerová, E. The Project Class in Central Europe: The Czech and Hungarian Cases. *Sociol. Rural.* **2006**, *46*, 3–21. [CrossRef]

environmental sciences proceedings

MDPI

Proceeding Paper

Environmental Citizenship of Students of Primary Education of a Greek University †

Marianna Kalaitsidaki * and Efthimia Baltsioti

Department of Primary Education, University of Crete, Gallou Campus, 74100 Rethimno, Greece; efthimaki111995@gmail.com

* Correspondence: mkalaitz@edc.uoc.gr; Tel.: +30-694-442-4082

† Presented at the 2nd International Conference of International Researchers of the Education for Environmental Citizenship 2022, 10–11 March 2022. Available online: https://enec-cost.eu/ireec22/.

Abstract: Exploring environmental citizenship is very important in the context of environmental education and sustainability. Global environmental problems require citizens who are active, critical-minded and able to act as agents of change for society and the environment. Environmental citizenship was defined for the first time with the contributions of many experts in the framework of the COST-Enec network. This paper examines the level of environmental citizenship of undergraduate students of the Department of Primary Education at a Greek university with a recently published questionnaire. Findings regarding the correlation with demographics (gender, year of study and place of origin (urban/rural) are reported.

Keywords: environmental citizenship; education for environmental citizenship; pro-environmental behavior; environmental education; university

Citation: Kalaitsidaki, M.; Baltsioti, E. Environmental Citizenship of Students of Primary Education of a Greek University. *Environ. Sci. Proc.* **2022**, *14*, 15. https://doi.org/10.3390/environsciproc2022014015

Academic Editors: Andreas Ch. Hadjichambis, Pedro Reis, Marie-Christine Knippels, Audronė Telesiene, Daphne Goldman, Demetra Paraskeva-Hadjichambi, Jan Cincera and Kateřina Jančaříková

Published: 11 March 2022

1. Introduction

Environmental citizenship is a basic concept of the vision of sustainable development for a sustainable world. The European COST ENEC Network for Environmental Citizenship, a network of over 130 scholars defined environmental citizenship, for the first time, as the 'responsible environmentally friendly behavior of citizens who act and contribute in society as agents of change, within the private or public sphere, in a local, national or international context, with collective actions, in order to help solving contemporary environmental problems, prevent new environmental problems from arising, as well as for achieving sustainability and a healthy relationship with nature' [1]. The aim of the present study is to examine the environmental citizenship of undergraduate primary school teachers of the University of Crete, and correlate this with demographics, such as gender, length of studies and place of origin (urban and rural).

1.1. Environmental Citizenship and Primary Education

The education of environmental citizenship requires a model, which can be applied to all educational levels and contexts with the appropriate differentiation. The educational approach to environmental citizenship (EEC)was developed in the framework of the ENEC network and six stages were proposed: inquiry, planning of action, civic participation, networking and sharing in scales, reinforcement of environmental and social change, assessment and reflection [2]. The successful application of educational interventions for the promotion of environmental citizenship in primary schools requires suitable approaches and a specially trained teacher [3].

1.2. Environmental Citizenship and Environmental Education

According to EEC's definition, there are eight outcomes, which can be achieved through actions in two dimensions (personal and collective), applied in two different spheres (private and public) and on three different scales (local, national and international). Based on this model of education for environmental citizenship, a tool named the Environmental Citizenship Questionnaire (ECQ) was developed by Hadjichambi and Paraskeva-Hadjichambi [4] to measure environmental citizenship.

The actions of environmental citizenship are recognized as actions in the public sphere when they affect the relations within the societies, and as actions in the private sphere when they affect the relationships between persons and societies [2]. HadjichambisandParaskeva-Hadjichambi [4] used the ECQ they developed to evaluate the environmental citizenship of 520 10th grade students in Cyprus. Their results showed that the students had high scores regarding their knowledge on environmental citizens, as well as on their past actions. However, their mean values regarding the attitudes and values of environmental citizens were quite low, and their values regarding skills and concepts about the environment were average.

2. Materials and Methods

In the present quantitative study, we used the ECQ questionnaire developed by Hadjichambis and Paraskeva-Hadjichambi [4]. The questionnaire consists of 76 Likert-type questions, separated in three different sections, all related to environmental citizenship. The first section investigates past and present actions which were or are carried out by environmental citizens (6). The second section regards participants' knowledge of environmental citizenship (11), opinions about environmental citizenship (12), skills of the environmental citizen (6), and the behavior (8) and values of the environmental citizens (15). The last section investigates future actions as environmental citizens: within the university campus (4), outside the campus (11) and as an agent of change (3) Demographic data regarding gender, semester of study and place of origin (urban/rural) were included. The questionnaire was pilot-tested with post-graduate students. Completion time was reported to be 15 min. small changes were required to be made. Because of the lockdown due to the COVID 19 pandemic, the questionnaire was sent in June 2021, via Google Forms, to the university emails of 994 students of Primary Education from years 1–5. The department has a 4-year study program but there some students that do not graduate on time and extend their study. Factor analysis was performed according to Hadjichambi and Paraskeva-Hadjichambi [4] the researchers that developed the questionnaire.

3. Results

In total 167 students responded, 146 women and 21 men. Regarding the semester of study, 4 students were from the first semester, 45 from the second, 1 from the third, 31 from the fourth, 4 from the fifth, and 43 from the sixth and above semesters 39. We tested differences between the males and females and the urban/rural place of origin of the students, based on the results of the ECQ factors and the nonparametric Mann-Whitney test, because of the abnormality of the dependent variables. No statistically significant difference was found between the two genders, nor between the urban/rural place of origin. We tested the differences between the years of study (Table 1), based on the results of the ECQ factors and the nonparametric Kruskal-Wallis test, because of the abnormality of the dependent variables. A statistically significant difference was found between the different years of study on Factor4, Skills of ECn: third-year students and above presented with a larger score than first- and second-year students (χ^2 = 10,678, p = 0.014).

Table 1. Testing of differences between years of study and ECQ nine factors.

Year		F1 Past Actions as ECn	F2 Knowledge for ECn	F3 Conceptions for ECn	F4 Skills of ECn	F5 Attitudes of ECn	F6 Values of ECn	F7 Future Actions inside School	F8 Future Actions Outside School	F9 Agents of Change
1st year	N	49	49	49	49	49	49	49	49	49
	Median	2.8571	2.4545	3.4167	2.5	3.5556	3.3529	3	2.5455	3.3333
	Range	1.57	2.82	1.92	2.33	1.44	1.88	2.75	2.18	1.67
2nd year	N	32	32	32	32	32	32	32	32	32
	Median	2.8571	2.3636	3.2917	2.5	3.5556	3.3529	2.75	2.6364	3.3333
	Range	1.29	2.18	1.42	2	1.33	1.41	2.75	2	1.67
3rd year	N	47	47	47	47	47	47	47	47	47
	Median	2.8571	2.6364	3.25	**2.8333**	3.4444	3.4118	2.75	2.6364	3.3333
	Range	1.43	2.55	2	2.83	1.56	1.18	2.5	2	1.67
4th year+	N	39	39	39	39	39	39	39	39	39
	Median	2.8571	2.6364	3.25	**2.8333**	3.5556	3.4118	3	2.6364	3.6667
	Range	1.29	2.55	1.58	1.83	1	1.24	2.5	2.27	2
Kruskal-Wallis	χ^2	1.095	7.239	0.624	10.678	1.106	4.724	2.102	2.51	0.076
	Asymp. Sig	0.778	0.065	0.891	**0.014**	0.776	0.193	0.552	0.473	0.995

4. Discussion

Our results indicate that during the 4-year program of study in the Department of Primary Education of the University of Crete, students have opportunities to develop the appropriate skills of an environmental citizen during the third and fourth years of their studies, but this is not the case for the other factors that define environmental citizenship. Therefore, more emphasis should be placed on the program of study to provide opportunities in the other areas that are of interest to environmental citizenship. It would be interest to see similar studies from students from other departments and/or other countries.

Author Contributions: Conceptualization, M.K.; methodology, M.K. and E.B.; formal analysis, E.B.; writing—original draft preparation, E.B.; Writing—review and editing, M.K. All authors have read and agreed to the published version of the manuscript.

Funding: The research received no external funding.

Institutional Review Board Statement: This study was conducted in accordance with the Declaration of Helsinki and it did not require ethical approval, The link to a digital questionnaire was sent by email to students' departmental emails that contain matriculation numbers, not names. The participation to the study was on a voluntary basis. Questionnaires were answered anonymously.

Informed Consent Statement: Not applicable. The students had the right to choose not to answer the survey.

Data Availability Statement: Not applicable.

Conflicts of Interest: The authors declare no conflict of interest.

References

1. European Network for Environmental Citizenship—ENEC. Defining Environmental Citizenship. 2018. Available online: http://enec-cost.eu/our-approach/enec-environmental-citizenship/ (accessed on 2 November 2020).
2. Hadjichambis, A.C.; Paraskeva-Hadjichambi, D. Education for Environmental Citizenship: The Pedagogical Approach. In *Conceptualizing Environmental Citizenship for 21st Century Education*; Hadjichambis, A.C., Reis, P., Paraskeva-Hadjichambi, D., Činčera, J., de Pauw, J., Gericke, N., Knippels, M.C., Eds.; Environmental Discourses in Science Education; Springer: Cham, Switzerland, 2020; pp. 237–261.

3. Činčera, J.; Romero-Ariza, M.; Zabric, M.; Kalaitzidaki, M.; Diez Bedmar, M. Environmental Citizenship in Primary Formal Education. In *Conceptualizing Environmental Citizenship for 21st Century Education*; Hadjichambis, A.C., Reis, P., Paraskeva-Hadjichambi, D., Činčera, J., de Pauw, J., Gericke, N., Knippels, M.C., Eds.; Environmental Discourses in Science Education; Springer: Cham, Switzerland, 2020; pp. 163–177.
4. Hadjichambis, A.C.; Paraskeva-Hadjichambi, D. Environmental Citizenship Questionnaire (ECQ): The Development and Validation of an Evaluation Instrument for Secondary School Students. *Sustainability* **2020**, *12*, 821. [CrossRef]

environmental sciences proceedings

MDPI

Proceeding Paper

A Liable Gender Approach in Environmental Grind in Albania †

Nevila Xhindi [1,*] and Enis Sokoli [2]

[1] Department of Public Administration, Mediterranean University of Albania, 1023 Tirana, Albania
[2] Program's Development, Center for Comparative and International Studies, 1023 Tirana, Albania; enissokoli@gmail.com
* Correspondence: nevila.xhindi@umsh.edu.al; Tel.: +355-682056910

† Presented at the 2nd International Conference of International Researchers of the Education for Environmental Citizenship 2022, 10–11 March 2022. Available online: https://enec-cost.eu/ireec22/.

Abstract: Albania has only begun to address environmental and sustainable development issues as recently as the 1990s. The environment has received the attention of consecutive governments that drafted the relevant legislation based on EU experience, ratified several conventions, and signed international agreements. Meanwhile, the national environment strategy and plan were approved, while environmental and sustainable development elements became cross-cutting issues. A particular aspect in the environmental legal framework is the gender aspect. This is relevant in contemporary society due to climate change, land degradation, and biodiversity conservation. All of these require gender attention and new environmental planning methods. This paper aims to analyze the implementation of the environmental legal framework through the gender perspective and gender roles.

Keywords: development; environment; women; legislation

Citation: Xhindi, N.; Sokoli, E. A Liable Gender Approach in Environmental Grind in Albania. *Environ. Sci. Proc.* **2022**, *14*, 16. https://doi.org/10.3390/environsciproc2022014016

Academic Editors: Andreas Ch. Hadjichambis, Pedro Reis, Marie-Christine Knippels, Audronė Telesiene, Daphne Goldman, Demetra Paraskeva-Hadjichambi, Jan Cincera and Kateřina Jančaříková

Published: 14 March 2022

1. Introduction

Scholars point out that women have a 'special' relationship with the environment, as active users or managers. In studies on development issues, this approach has become known as WED (Women, Environment, and Development) [1,2]. WED's theoretical and well-known discussions focus on the current roles of women [3] as users and managers of natural resources [4,5] and, of course, the impact this has on society and the environment. As a common emphasis of policy and intervention, it represents an apparent attempt to link previous approaches to WID (Women in Development) with the latest environmental policy concerns.

Albania has a set of already consolidated laws which focus on the environment or treat it as a cross-sectoral issue. National environmental legislation has been developed in full compliance with the European legal framework (Acquis Communitaire) because integration into the European Union is the country's principal strategic and political objective. Additionally, during the last 15 years, both the legal and political framework has been agreed with the Sustainable Development Goals (SDG), which is critical to ensure that the new laws in Albania are in line with the principles of sustainable development. In the meantime, the gender aspect, especially the empowerment of women and gender equality [6–13], is present in all laws, sectorial or cross-sectoral strategies, ratified conventions, or signed agreements related to the environment and development.

Regarding the structure, this paper presents the theoretical framework on gender and environment, a brief description of the environmental situation in Albania, with a focus on gender mainstreaming in environmental policies, and a scientific perspective towards the connection between gender and biodiversity loss desertification. Climate change impacts all this in the local context of equal participation of women and men in the planning, implementing, and monitoring of various environmental issues.

2. Methodology

The methodology used is based mainly on a desk review of existing reports, research, strategies, the legislative framework in Albania, and different international policies related to the environment analyzed from a gender perspective. The international experience presented is a starting point for recommendations on how it can be adopted in the Albanian context. The focus is on recognizing the multiple dimensions of and interactions between gender equality and the environment within an integrated policy framework, considering inclusive growth and environmental considerations at local, national, and international levels.

3. Conclusions

Embarking on a gender-mainstreaming strategy calls for knowledge and understanding of the issue and validation of women's contributions to sustainable development by ensuring full and active participation of women together with gender equity in natural resource management, research, planning, and decision making at all levels.

Gender mainstreaming in environmental policies is an issue for both women and men. Therefore, the strategy to encourage their equal participation deliberately addresses men and makes them—together with women—responsible for a gender approach to environmental work. Additionally, the focus on gender must not only exist on the international level, still, it must advance within a specific local context, considering other elements of social gender diversity such as age, skills, ethnicity, etc.

4. Recommendations

The engagement of women and men in participatory processes at the local level is not a new concept in the Albanian context; many initiatives have been undertaken to date, especially regarding women's engagement in planning and decision-making processes. This engagement is crucial, not simply in the planning phase such as drafting the local development plans in the frame of decentralization and the new territorial reform, but also during the implementation and monitoring phase. These experiences should further strengthen women's voices in planning, implementing, and monitoring different environmental policies and actions

Author Contributions: Conceptualization, N.X. and E.S.; methodology, N.X.; formal analysis, N.X.; investigation, E.S.; data curation, E.S.; writing—original draft preparation, N.X.; writing—review and editing, N.X. All authors have read and agreed to the published version of the manuscript.

Funding: This research received no external funding.

Institutional Review Board Statement: Not applicable.

Informed Consent Statement: Not applicable.

Data Availability Statement: No new data were created or analyzed in this study.

Acknowledgments: This study is partly inspired by the ENEC Cost Action European Networks for Environmental Citizenship—CA16229 supported by COST (European Cooperation in Science and Technology—Horizon 2020).

Conflicts of Interest: The authors declare no conflict of interest.

References

1. Pandey, S. Women, environment, and sustainable development. *Int. Soc. Work.* **1998**, *4*, 339–355. [CrossRef]
2. Koroleva, A. Chapter 11: Women and Environment. In *Global Women's Issues: Women in the World Today, Extended Version*; Bureau of International Information Programs, United States Department of State: Washington, DC, USA, 2001.
3. Jackson, C. *Doing What Comes Naturally? Women and Environment in Development*; Elsevier Ltd.: Norwich, UK, 1993.
4. Vakoch, D.A.; Mickey, S. *Women and Nature? Beyond Dualism in Gender, Body, and Environment*; Rutledge: London, UK, 2019.

5. Davidson, J. Women's relationship with the environment. *Gend. Dev.* **1993**, *1*, 5–10. [CrossRef] [PubMed]
6. IRG an RTI Company. Energy Strategy for Albania 2017–2030. Available online: https://www.h2o-initiative.org/wp-content/uploads/documents-public/Albania/USAID-2018-Albanian-Energy-Strategy.pdf (accessed on 12 September 2021).
7. Ministry of Agriculture and Rural Development. Cross-Sectorial Strategy for Rural and Agricultural Development. Available online: https://bujqesia.gov.al/wp-content/uploads/2018/02/STRATEGJIA_NDERSEKTORIALE.pdf (accessed on 13 October 2021).
8. Council of Ministers and Minister of State for Protection of Enterprise. Strategy for Interaction between the Enterprise and Public Institutions. Available online: https://sipermarrja.gov.al/strategjia-per-nderveprim-mes-sipermarrjes-dhe-institucioneve-publike (accessed on 4 November 2021).
9. Ministry of Infrastructure and Energy. Sectorial Strategy of Transport and Action Plan 2016–2020. Available online: https://www.infrastruktura.gov.al/wp-content/uploads/2020/07/3rd-Monitoring-Report-of-Sectorial-Transport-Strategy-and-Action-Plan-2016-2020_June-2020.pdf (accessed on 4 November 2021).
10. Ministry of Tourism and Environment. National Tourism Strategy 2019–2023. Available online: https://turizmi.gov.al/wp-content/uploads/2019/12/National-Tourism-Strategy-2019-2023-EN.pdf (accessed on 20 October 2021).
11. Ministry of Economy, Trade and Energy. Cross Sector Strategy for Regional Development. Available online: Shtetiweb.org/wp-content/uploads/2014/05/rdcs_final_document.pdf (accessed on 3 November 2021).
12. Ministry of Health and Social Protection. National Strategy for Social Protection 2015–2020. Available online: http://shendetesia.gov.al/wp-content/uploads/2018/06/Strategjia_Kombetare_per_Mbrojtjen_Sociale_2015-2020.pdf (accessed on 13 October 2021).
13. National Strategy for Sustainable Tourism Development 2019–2023 in the Framework of Coordination of Measures for Sustainable Development. Available online: http://www.mjedisi.gov.al/wp-content/uploads/2019/06/Strategjia-Komb%C3%ABtare-e-Turizmit-2019-2023.pdf (accessed on 23 October 2021).

environmental sciences proceedings

MDPI

Proceeding Paper

Construct of Youth Environmental Citizenship among International Large-Scale Educational Studies †

Saiki Lucy Cheah [1,*] and Lihong Huang [2]

1 Helsinki Institute of Sustainability Science (HELSUS), University of Helsinki, 00100 Helsinki, Finland
2 Norwegian Social Research—NOVA, Oslo Metropolitan University, 0130 Oslo, Norway; lhuang@oslomet.no
* Correspondence: saiki.cheah@helsinki.fi

† Presented at the 2nd International Conference of International Researchers of the Education for Environmental Citizenship 2022, 10–11 March 2022. Available online: https://enec-cost.eu/ireec22/.

Abstract: Environmental citizenship has become an integral element of civic and citizenship education curricula, both globally and at the European level. Previous analysis provided important evidence that environmental-citizenship education practice in Nordic schools has a significant positive association with heightened attitudes and magnified behaviours among students toward environmental actions now and in the future. This paper starts with critiques on the strengths and limitations of ICCS 2016 data measures of youth environmental citizenship, and then it provides an overview of measures applied in a few other large-scale international studies such as PISA 2015 and PISA 2018. At the end, we present a measurement proposal on environmental citizenship through combining the measures from these studies.

Keywords: environmental citizenship; Nordic countries; citizenship education; International Civic and Citizenship Education Study (ICCS); Programme for International Student Assessment (PISA)

Citation: Cheah, S.L.; Huang, L. Construct of Youth Environmental Citizenship among International Large-Scale Educational Studies. *Environ. Sci. Proc.* **2022**, *14*, 17. https://doi.org/10.3390/environsciproc2022014017

Academic Editors: Andreas Ch. Hadjichambis, Pedro Reis, Marie-Christine Knippels, Audronė Telesiene, Daphne Goldman, Demetra Paraskeva-Hadjichambi, Jan Cincera and Kateřina Jančaříková

Published: 16 March 2022

1. Introduction

Environmental citizenship has become an integral element of civic and citizenship education curricula, both globally and at the European level [1,2]. Previous analysis [3,4] provided results of measures using students' concerns, values, engagement, and intended future participation on environmental issues to construct environmental citizenship and its relation to student home socioeconomic background, gender, and migrant status. The results show that there are both similarities and small variations in elements of student environmental citizenship among the Nordic countries and in comparison with their European and international peers. Nordic students stand out as concerned environmental citizens while they rank somehow lower than their European and international peers in engagement, values, and intended participation of environmental citizenship. We find that student environmental citizenship is socially divided in all Nordic countries as it differs significantly between students from different socioeconomic strata and genders. Although not all differences of student environmental citizenship by migrant status are statistically significant among the Nordic countries, we find some significant influence of migrant status interaction with socioeconomic status and genders.

We follow a recent and comprehensive definition of environmental citizenship by the European Network for Environmental Citizenship (ENEC, 2017–2022):

Environmental citizenship is the responsible pro-environmental behavior of citizens who act and participate in society as agents of change in the private and public sphere on a local, national, and global scale, through individual and collective actions, in the direction of solving contemporary environmental problems, preventing the creation of new environmental problems, and achieving sustainability and developing a healthy relationship with nature (http://enec-cost.eu/our-approach/) (assessed on 16 March 2022 ENEC, 2018) [5].

The ENEC definition of Environmental citizenship includes elements of knowledge, skills, attitudes, values. These elements are also compatible with the UNESCO definition of Education for Sustainable Development (ESD), "Education for Sustainable Development Education for Sustainable Development (ESD) empowers learners with knowledge, skills, values and attitudes to take informed decisions and make responsible actions for environmental integrity, economic viability and a just society." [6].

2. Existing Data and Measures of Environmental Citizenship

We present the variables of interest measuring student concerns, values, attitudes, behaviors and intended behaviors related to environmental issues from existing data of ICCS 2016 in Table 1 and PISA 2016 and PISA 2018 in Table 2. Table 1 shows students' responses to the measurement items in percentages of Europe comparing with those of international. The preliminary results in Table 1 indicate an overall similarity between European students and students of the world. In Table 2, without data analysis, we only present the measures of variables relevant for environmental citizenship existing in PISA 2015 and PISA 2018 studies, to serve the purpose of our conference presentation of proposing a measurement of environmental citizenship.

Table 1. Description of measures from International Civic and Citizenship Education Study (ICCS) 2016 study.

Items of Interest in ICCS 2016	Response Alternative	% Europe	% International
Have you ever been involved in environmental organizations or groups (STQ15b)	Yes, have been	25.1 (0.3)	28.1 (0.2)
	Never	74.9 (0.3)	71.9 (0.2)
	Total	100	100
Have you evern been participating in an activity to make the school more environmentally friendly (STQ16f)	Yes, have done	43.5 (0.4)	49 (0.3)
	Never	56.5 (0.4)	51 (0.3)
	Total	100	100
A school, to what extent have you learnt how to protect the environment (STQ18c)	A lot	41.2 (0.4)	46.1 (0.3)
	To some extent	37.9 (0.3)	34.7 (0.2)
	A little	16.5 (0.2)	15.2 (0.2)
	Not at all	4.4 (0.2)	3.9 (0.1)
	Total	100	100
How important for a good adult citizen to participate in protecting the environment (STQ23j)	Very important	38.2 (0.3)	44.1 (0.2)
	Quite important	43.8 (0.3)	41.3 (0.2)
	Not very important	15.1 (0.2)	12.1 (0.1)
	Not important at all	2.9 (0.1)	2.5 (0.1)
	Total	100	100
How important for a good adult citizen make personal effort to protect the environment (STQ23n)	Very important	43.1 (0.3)	48.7 (0.2)
	Quite important	43.9 (0.3)	40.7 (0.2)
	Not very important	10.9 (0.2)	8.8 (0.1)
	Not important at all	2.1 (0.1)	1.9 (0.1)
	Total	100	100
Biggest threat to the world future is pollution (STQ28a)	To a large extent	73.6 (0.3)	75.3 (0.2)
	To a moderate extent	21.5 (0.3)	20 (0.2)
	To a small extent	3.7 (0.1)	3.5 (0.1)
	Not at all	1.2 (0.1)	1.2 (0.0)
	Total	100	100

Table 1. *Cont.*

Items of Interest in ICCS 2016	Response Alternative	% Europe	% International
Biggest threat to the world future is climate change (STQ28i)	To a large extent	55.6 (0.3)	56.8 (0.2)
	To a moderate extent	29.6 (0.3)	28.4 (0.2)
	To a small extent	11.9 (0.2)	11.9 (0.1)
	Not at all	2.9 (0.1)	2.8 (0.1)
	Total	100	100
When becoming adult, will you make personal efforts to protect the environment (STQ31j)	I would certainly do	29.6 (0.3)	38 (0.2)
	I would probably do	45 (0.3)	41.4 (0.2)
	I would probably not do	19.2 (0.2)	15.3 (0.2)
	I would certainly not do	6.1 (0.2)	5.3 (0.1)
	Total	100	100

Source: Table A1 in [4].

Table 2. Measures from PISA 2015 and 2018 studies.

Items of Interest in PISA 2015	Response Alternatives
Students' responses to question "How informed are you about the following environmental issues?" (ST092)	1 = I have never heard of this 2 = have heard about this but I would not be able to explain what it is really about 3 = I know something about this and could explain the general issue 4 = I am familiar with this and I would be able to explain this well
The increase of greenhouse gases in the atmosphere	
The use of genetically modified organisms (<GMO>)	
Nuclear waste	
The consequences of clearing forests for other land use	
Air pollution	
Extinction of plants and animals	
Water shortage	
Students' responses to question "Do you think problems associated with the environmental issues below will improve or get worse over the next 20 years?" (ST093)	1 = improve 2 = stay about the same 3 = get worse
Air pollution	
Extinction of plants and animals	
Clearing of forests for other land use	
Water shortages	
Nuclear waste	
The increase of greenhouse gases in the atmosphere	
The use of genetically modified organisms (GMO)	
Students' responses to question "How easy do you think it would be for you to perform the following tasks on your own? (ST129)	
Predict how changes to an environment will affect the survival of certain species	1 = I couldn't do this, 2 = I would struggle to do this on my own, 3 = I could do this with a bit of effort, 4 = I could do this easily
Students' responses to question "How often do you do these things?" (ST146)	
Follow news of science, environmental, or ecology organisations via blogs and microblogging	1 = very often 2 = regularly 3 = sometimes 4 = never or hardly ever

Table 2. *Cont.*

Items of Interest in PISA 2018	Response Alternatives
Students' responses to questions "How easy do you think it would be for you to perform the following tasks on your own?" (ST196)	
Explain how carbon-dioxide emissions affect global climate change	1 = I couldn't do this, 2 = I would struggle to do this on my own, 3 = I could do this with a bit of effort, 4 = I could do this easily
Establish a connection between prices of textiles and working conditions in the countries of production	
Discuss the different reasons why people become refugees	
Explain why some countries suffer more from global climate change than others	
Explain how economic crisis in single countries affect the global economy	
Discuss the consequences of economic development on the environment	
Students' responses to question "How informed are you about the following topics?" (ST197)	
ST197Q01HA: Climate change and global warming	1 = I have never heard of this, 2 = I have heard about this but I would not be able to explain what it is really about, 3 = I know something about this and could explain the general issues, 4 = I am familiar with this and I would be able to explain this well
Students' responses to question "Are you involved in the following activities?" (ST222)	
I reduce the energy I use at home (e.g., by turning the heating down or turning the air conditioning up or down or by turning the lights when leaving a room) to protect the environment	Yes = 1 No = 2
I choose certain products for ethical or environmental reasons, even if they are a bit more expensive	
I sign environmental or social petition online	
I keep myself informed about world events via Twitter or Facebook	
I boycott products or companies for political, ethical or environmental reasons	
I participate in activities promoting equality between men and women	
I participate in activities in favour of environmental protection	
I regularly read websites on international social issues (e.g., poverty, human rights)	
Students' responses to question "To what extent do you agree with the following statements?" (ST219)	
ST219Q06HA: Looking after the global environment is important to me	1 = Strongly disagree, 2 = Disagree, 3 = Agree, 4 = Strongly agree

Source: Student Questionnaire for Pisa 2018 www.oecd.org/pisa/data (accessed on 16 March 2022).

3. Discussion and Conclusions

There have been studies that explore concepts and measures relevant to environmental education; in this paper we have categorised environmental-related questions formulated in the ICCS 2016 and PISA studies into several environmental citizenship attributes. First, the ICCS study appears to focus on students' concerns, values and future intentions in/outside of school engagement that are interpreted as environmental citizenship attributes. Second, the PISA studies seem to focus on knowledge, skills, engagement, dialogue, social media participation, and habit formation that are interpreted as environmental citizenship attributes. Nonetheless, existing studies such as teachers' understanding of sustainable development and student attitudes and consciousness towards the environment, school education and student learning for environmental citizenship, and a collective effort in conceptualization of environmental citizenship [7] also explore concepts and contribute to measures of environmental citizenship. However, the concepts and measures of environmental citizenship are rather fragmented geographically and socio-culturally, as well as across different scientific disciplines at different analytical levels. Hence, this array of studies may also contribute to a current lack of comprehensive measurement and perhaps enable researchers to assess and compare youth environmental citizenship across systems and national borders. This preliminary analysis of the ICCS and PISA studies has led to our proposing a new measurement including eight items measuring students' concerns, values, and future intentions in/outside of school engagement, and knowledge, skills, dialogue, social media participation and habit formation that are conducive to pro-environmental movements and change.

Author Contributions: Conceptualization, S.L.C. and L.H.; methodology, L.H.; formal analysis, S.L.C. and L.H.; investigation, S.L.C.; resources, L.H.; writing—original draft preparation, S.L.C. and L.H.; writing—review and editing, S.L.C.; visualization, L.H. All authors have read and agreed to the published version of the manuscript.

Funding: This research did not receive any external funding.

Institutional Review Board Statement: Not applicable.

Informed Consent Statement: Not applicable.

Data Availability Statement: Not applicable.

Conflicts of Interest: The authors declare no conflict of interest.

References

1. Council of Europe. *Reference Framework of Competences for Democratic Cultures*; Council of Europe Publishing: Strasbourg, France, 2018.
2. Gericke, N.; Huang, L.; Knippels, M.C.; Christodoulou, A.; Van Dam, F.; Gasparovic, S. Environmental citizenship in secondary formal education: The importance of curriculum and subject teachers. In *Conceptualizing Environmental Citizenship for 21st Century Education*; Hadjichambis, A., Reis, P., Paraskeva-Hadjichambi, D., Činčera, J., Pauw, J., Gericke, N., Knippels, M., Eds.; Environmental Discourses in Science Education; Springer: Cham, Switzerland, 2020; Volume 4. [CrossRef]
3. Cheah, S.L.; Huang, L. Environmental citizenship in a Nordic and citizenship education context. *Nord. J. Comp. Int. Educ.* **2019**, 3, 88–104. Available online: https://journals.hioa.no/index.php/nordiccie/article/view/3268/3383 (accessed on 1 March 2022). [CrossRef]
4. Huang, L.; Cheah, S.L. The young environmental citizens in Nordic Countries: Their concerns, values, engagement, and intended future actions. In *Northern Lights on Civic and Citizenship Education: A Cross-National Comparison of Nordic Data from ICCS*; Biseth, H., Hoskins, B., Huang, L., Eds.; SpringerOpen: Cham, Switzerland, 2021; Chapter 6; pp. 123–146.
5. European Network of Environmental Citizenship (ENEC). Defining Environmental Citizenship. 2018. Available online: http://enec-cost.eu/our-approach/enec-environmental-citizenship/ (accessed on 1 March 2022).
6. UNESCO. What Is Education for Sustainable Development? Available online: https://en.unesco.org/themes/education-sustainable-development/what-is-esd (accessed on 1 March 2022).
7. Hadjichambis, A.C.; Reis, P.; Paraskeva-Hadjichambi, D.; Činčera, J.; Boeve-de Pauw, J.; Gericke, N.; Knippels, M.-C. (Eds.) *Conceptualizing Environmental Citizenship for 21st Century Education*; Environmental Discourses in Science Education; Springer: Cham, Switzerland, 2020; Volume 4, Available online: https://link.springer.com/book/10.1007/978-3-030-20249-1 (accessed on 1 March 2022).

environmental sciences proceedings

MDPI

Proceeding Paper

Video Annotations for the Development of Environmental Citizenship during Initial Teacher Education †

Daniel Cebrián-Robles [1,*], Enrique España-Ramos [1] and Pedro Reis [2]

1 Science Education Department, University of Málaga, 29016 Málaga, Spain; enrienri@uma.es
2 Instituto de Educação, Universidade de Lisboa, 1649-013 Lisboa, Portugal; preis@ie.ulisboa.pt
* Correspondence: dcebrian@uma.es

† Presented at the 2nd International Conference of International Researchers of the Education for Environmental Citizenship 2022, 10–11 March 2022; Available online: https://enec-cost.eu/ireec22/.

Abstract: The social and environmental issues facing society call for changes in educational approaches. The massive use of video as a resource for dissemination on social networks calls for further analysis and reflection in education. A video analysis activity using annotations with the CoAnnotation.com platform to develop environmental citizenship is presented. In total, 104 preservice elementary teachers participated in work at Malaga University (Spain). The examined socio-scientific issue was related to illegal mining in Venezuela. Students were able to identify environmental, health, and cultural issues, in this order. The video annotation helped discuss and map a socio-scientific problem to facilitate the analysis of its complexity.

Keywords: video annotation; socio-scientific issue; preservice primary teachers

Citation: Cebrián-Robles, D.; España-Ramos, E.; Reis, P. Video Annotations for the Development of Environmental Citizenship during Initial Teacher Education. *Environ. Sci. Proc.* **2022**, *14*, 18. https://doi.org/10.3390/environsciproc2022014018

Academic Editors: Andreas Ch. Hadjichambis, Marie-Christine Knippels, Audronė Telesiene, Daphne Goldman, Demetra Paraskeva-Hadjichambi, Jan Cincera and Kateřina Jančaříková

Published: 21 March 2022

1. Introduction

The complexity of the important social and environmental issues of the current global crisis poses new educational challenges that require special attention. Among them, we highlight the challenge of facilitating citizens' empowerment for democratic involvement in social issues related to science and technology through socio-political activism [1]. Students' socio-political action, according to Bencze and Carter [2], has the potential to improve: (a) their knowledge and awareness of these topics; (b) their research and citizenship skills; and, finally, (c) the well-being of individuals, societies, and environments. However, in order to develop these competencies in students, activism must first be introduced into teacher training. Indeed, preservice teachers will only be able to engage their students in activism projects to change society and the environment if they engage in socio-political actions themselves [3]. One of the methods that can carry out these types of social transformations is video podcasting or vodcasting [4], mainly if the videos are produced and distributed by students as a form of socio-political action [5].

With this aim, in the context of science education, we want to contribute to the training of environmental citizens, agents of change capable of solving contemporary environmental issues through individual and collective actions [6]. To this end, this paper proposes video annotations as a teaching activity for the analysis and discussion of activist videos with socio-scientific content.

2. Methodology

2.1. Participants

This teaching activity represented a preliminary step in the development of an educational activism programme with 104 preservice elementary teachers (PSETs) (aged between 20 and 21 years) at Malaga University (Spain).

2.2. Video Annotation Activity

The activity consisted of viewing an activist video about illegal mining in the Venezuelan Amazon and analysing it through collaborative online annotation of fragments of the video [7]. With this purpose, the open-access platform CoAnnotation.com was used to select fragments of the video in which the students were told to do the following.

(1) Choose one of the problems created by illegal mining mentioned in the video. Choose the one you believe is the most serious and justify your decision in the annotation. (Each student should have one annotation.)

(2) Choose the best solution from among the recommended solutions to the problem you have seen. If the video does not offer a solution to the problem you have selected, you can annotate a section of the video and submit your own. (Each student should have one annotation.)

(3) Identify sections of the video that, in your opinion, might be enhanced to increase public awareness regarding illegal mining in Venezuela. (Each student may add one or more annotations.)

Figure 1 shows a screenshot of the CoAnnotation platform with the task to be carried out, demonstrating the video and the visible statistics of where the annotations on the video are concentrated.

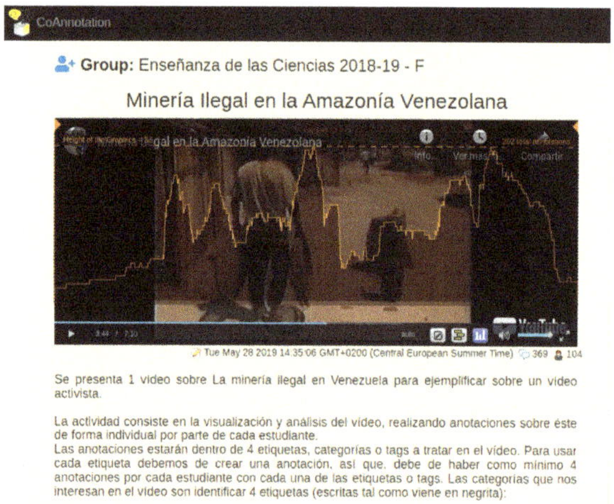

Figure 1. CoAnnotation platform with the statistics of the annotations.

Subsequently, a discussion session was held based on the analysis of the set of annotations on the different fragments of the video.

2.3. Data Collection and Analysis

For the content analysis of the video annotations and the class discussions' observation records, an expert-validated categorisation system was used. The data were read several times until agreement was reached between the authors, generating three main categories that coincide in both the problems and the solutions: environmental, health, and socio-cultural. An annotation could be categorised in more than one category.

A Sankey diagram was used to visualise the relationships between the problems and the solutions identified by the PSETs. The Sankey diagram shows the number of students who identify a particular problem on the left and the number of students who identify a solution on the right. With coloured lines, environmental solutions could be identified in green, socio-cultural solutions in blue, and health solutions in red. The thickness of the line size depended on the number of students who identified a specific problem and connected it to a particular solution.

3. Results

The 104 students contributed with 369 annotations to the video as a whole. Only the results of the annotations associated with the problem and specific solution are shown.

Figure 2 represents the number of students who identify each of the three types of problem (environmental, health, and socio-cultural), together with the kind of solution they propose.

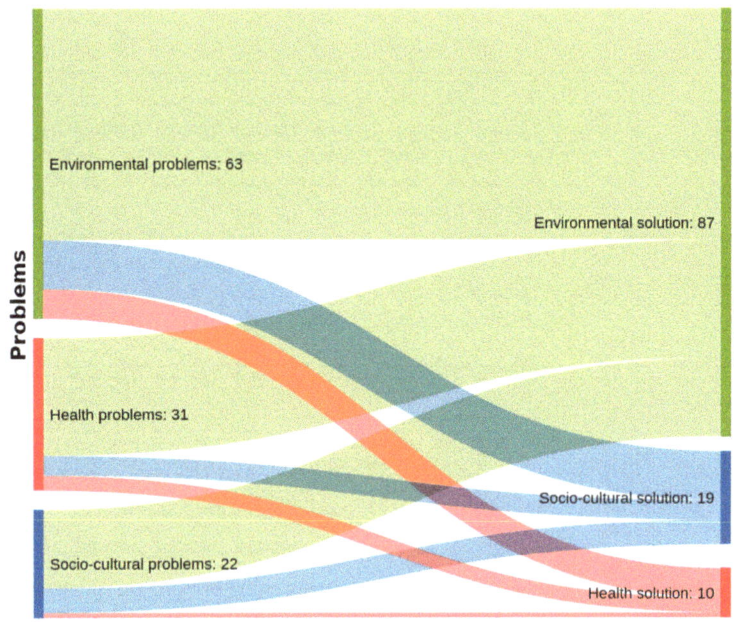

Figure 2. Sankey results of problem and solution annotations.

A large majority of students identified environmental problems (63 students), compared to health problems (31 students) and socio-cultural problems (22 students).

As one might expect, a solution for a problem of the same type is commonly advocated, such as an environmental solution for an environmental issue. Surprisingly, Figure 2 further shows that all three types of solutions were primarily proposed to address an environmental issue (the thickest line from each kind of problems links to the environmental solution). One student, for example, addressed the problem as follows: "This mining not only has an impact on the natural world but also on a cultural level, creating serious health problems, as well as a rise in prostitution, drug use, alcoholism, flight of capital and smuggling. This has a major social impact". Although this student mentioned health and societal problems, his solution was primarily focused on the environment: "The solution lies in building a people's task force to develop environmental education and communication initiatives so as to ensure that nature is fully respected and promote sustainable development".

4. Discussion and Conclusions

While some students mentioned all three types of impact (environmental, health, and social-cultural) in their annotations, their preferred solution was only focused on the environmental level. This leads us to believe that this activity could contribute to the development of environmental citizenship based on pro-environmentally responsible behaviours, in which citizens act and participate in society as change agents in the private and public spheres, at local, national, and global scales, through individual and collective actions aimed at solving current environmental problems, preventing the emergence of new environmental problems, and achieving environmental goals [4].

Video annotations enable the discussion mapping around socio-scientific problems, including an identification of those problems, participating citizen groups, possible causes-effects, and prospective solutions. This can be used in conjunction with other approaches, such as controversy mapping [8], which use actor–network theory to analyse the interconnections between actants [9]. The video and annotations are both helpful to teacher education since they complement Hodson [1]'s notion of learning about action, learning from action, and learning through action. Students in this study watched an activist video to learn about action and identify the skills and methods needed to make an activist video. As another task, they analysed the video's information to see if it was possible to strengthen the activist component. At this time, students were developing their skills in producing activist videos with a more significant impact on viewers. These skills can empower students as video creators and prospective activists, training them for action (for example, through social media distribution of videos) [10]. Preservice teachers can use video annotations to analyse the outcomes of activist projects they or others have undertaken, allowing them to learn from the action.

In conclusion, we consider that the use of video annotations can aid the development of environmental citizenship during the initial training of teachers, allowing them to learn a teaching resource adequate for the identification and critical analysis of important and complex socio-scientific problems: a preliminary step to socio-political action [1].

Author Contributions: Conceptualization, D.C.-R., E.E.-R. and P.R.; methodology, D.C.-R., E.E.-R. and P.R.; software, D.C.-R.; validation, D.C.-R., E.E.-R. and P.R.; formal analysis, D.C.-R.; investigation, D.C.-R.; resources, D.C.-R.; data curation, D.C.-R. and E.E.-R.; writing—original draft preparation, D.C.-R.; writing—review and editing, D.C.-R., E.E.-R. and P.R.; visualization, D.C.-R.; supervision, D.C.-R., E.E.-R. and P.R. All authors have read and agreed to the published version of the manuscript.

Funding: This research received no external funding.

Institutional Review Board Statement: Not applicable.

Informed Consent Statement: Informed consent was obtained from all subjects involved in the study.

Data Availability Statement: Not applicable.

Acknowledgments: This research was part of the Spanish National Plan R+D+I Project (grant number: PID2019-105765GA-I00).

Conflicts of Interest: The authors declare no conflict of interest.

References

1. Hodson, D. Going Beyond STS Education: Building a Curriculum for Sociopolitical Activism. *Can. J. Sci. Math. Technol. Educ.* **2021**, *20*, 592–622. [CrossRef]
2. Bencze, L.; Carter, L. Globalizing Students Acting for the Common Good. *J. Res. Sci. Teach.* **2011**, *48*, 648–669. [CrossRef]
3. Linhares, E.F.; Reis, P. Formar Futuros Professores Para a Ação Sociopolítica No Contexto Da Educação Em Ciências. *Rev. Bras. Ensino Ciência Tecnol.* **2018**, *11*, 86–103. [CrossRef]
4. Gkatzidou, S.; Pearson, E. Vodcasting: A Case Study in Adaptability to Meet Learners' Needs and Preferences. In Proceedings of the ICT: Providing Choices for Learners and Learning, Singapore, 2–5 December 2007; Citeseer: Singapore, 2007; pp. 325–332.
5. Marques, A.R.; Reis, P. Research-Based Collective Activism Through the Production and Dissemination of Vodcasts About Environmental Pollution in the 8th Grade. *Sisyphus-J. Educ.* **2017**, *5*, 116–137.
6. Hadjichambis, A.C.; Reis, P.; Paraskeva-Hadjichambi, D.; Činčera, J.; Boeve-de Pauw, J.; Gericke, N.; Knippels, M.-C. (Eds.) *Conceptualizing Environmental Citizenship for 21st Century Education*; Springer: Cham, Switzerland, 2020; ISBN 9783030202484.
7. Cebrián-Robles, D.; España-Ramos, E.; Reis, P. Introducing Preservice Primary Teachers to Socioscientific Activism through the Analysis and Discussion of Videos. *Int. J. Sci. Educ.* **2021**, *43*, 2457–2478. [CrossRef]
8. Latour, B. La Cartographie Des Controverses. *Technol. Rev.* **2007**, *2*, 82–83.
9. Latour, B. *Reassembling the Social: An Introduction to Actor-Network-Theory*; Oxford University Press: New York, NY, USA, 2005.
10. Cebrián-Robles, D.; España-Ramos, E.; Reis, P. Programa Formativo Sobre El Activismo Colectivo Basado En La Indagación Para La Formación Inicial Del Profesorado. In *Enseñanza de las Ciencias y Problemas Relevantes de la Ciudadanía. Transferencia al aula*; Cebrián-Robles, D., Franco-Mariscal, A.J., Lupión-Cobos, T., Acebal-Expósito, C., Blanco-López, Á., Eds.; Graó: Barcelona, Spain, 2021; pp. 223–237, ISBN 9788418058929.

environmental sciences proceedings

MDPI

Proceeding Paper

Education for Environmental Citizenship: A Master's Course Incorporating International Collaboration †

Ariel Sarid [1,*] and Daphne Goldman [2]

[1] M.Ed. Management and Organization of Educational Systems, Faculty of Education, Beit Berl College, Beit Berl 4490500, Israel

[2] Department of Environmental Science and Agriculture, Faculty of Education, Beit Berl College, Beit Berl 4490500, Israel; dafnag@beitberl.ac.il

* Correspondence: ariels@beitberl.ac.il

† Presented at the 2nd International Conference of International Researchers of the Education for Environmental Citizenship 2022, 10–11 March 2022. Available online: https://enec-cost.eu/ireec22/.

Abstract: Education for environmental citizenship (EEC) is a crucial component of the effort to promote societies that are aware of sustainable lifestyles and able and motivated to adopt them, and such education is acknowledged as a key component within the discourse around 21st-century global processes. A global trend calling for the incorporation of environmental and sustainability education in formal and nonformal educational systems is becoming increasingly prevalent. In line with this global trend, the present paper presents a course that will be conducted within the framework of the Master's programs at Beit Berl College in cooperation with international partners (lecturers, students). The course incorporates the ENEC's (COST action) conceptualization of EC and EEC, including the pedagogical model developed within ENEC. The course has the potential to have a wide impact on educational practice and curriculum in the field (both formal and non-formal).

Keywords: education for environmental citizenship; higher education; transformative change; social change agents

Citation: Sarid, A.; Goldman, D. Education for Environmental Citizenship: A Master's Course Incorporating International Collaboration. *Environ. Sci. Proc.* **2022**, *14*, 20. https://doi.org/10.3390/environsciproc2022014020

Academic Editors: Andreas Ch. Hadjichambis, Pedro Reis, Marie-Christine Knippels, Audronė Telesiene, Demetra Paraskeva-Hadjichambi, Jan Cincera and Kateřina Jančaříková

Published: 8 April 2022

Publisher's Note: MDPI stays neutral with regard to jurisdictional claims in published maps and institutional affiliations.

1. Introduction

Education for environmental citizenship (EEC) is a crucial component of the effort to promote societies that are aware of aware of sustainable lifestyles and able and motivated to adopt them, and such education is acknowledged as a key component within the discourse around 21st-century global processes. A global trend calling for the incorporation of environmental and sustainability education in formal and non-formal educational systems is becoming increasingly prevalent [1–3]. EEC is identified as an integral component of citizenship/civic education. Cao claims that education for citizenship and education on the environment are on a gradual path of convergence [4]. The ability to incorporate locally and culturally adapted EEC is essential for educators leading collaborative and participatory local or international initiatives.

The course we shall be presenting will be conducted within the framework of the Master's programs at Beit Berl College in cooperation with international partners (lecturers, students), utilizing partnerships created in ENEC [5] and beyond, thus realizing ENEC's goal to broaden its impact and apply the knowledge produced within the network.

2. Course Aims and Structure

The course incorporates the ENEC's (COST action) conceptualization of EC and EEC, including the pedagogical model developed within ENEC [6]. It focuses on EEC as part of the contemporary perspective on citizenship, including its role in advocating transformative change (at the individual, community, and social-system levels) by developing agents of social change.

The course aims to

- Develop a broad understanding of the crucial role of EEC in today's risk society.
- Acquire a broad, basic understanding of EC from theoretical and practical perspectives.
- Connect EC to education in general and to contemporary pedagogies.
- Develop educational interventions for cultivating EC in different educational and cultural contexts. By incorporating opportunities for cooperation and joint learning among different groups of students from different (cultural and educational) backgrounds, the course facilitates a deeper understanding of what it means to act as an educational agent of social change.

These aims are realized in the five units composing the course structure (see Table 1). Each unit concentrates on a different dimension of EEC, from understanding the importance of EEC in today's reality, through the examination of pedagogies accommodating EC, to the development of educational interventions that cultivate what it means to act as an educational agent of social change [7].

Learning and Teaching Methods

The course applies various teaching–learning methods that reflect pedagogical developments in Higher Education since the COVID-19 pandemic. Table 1 outlines the course contents and selected activities. The course will emphasize the active role of learners in the learning process, apply collaborative learning methods, and utilize digital technologies to ensure meaningful connections between learners coming from different socio-cultural backgrounds and academic disciplines. The use of ICTs in learning will also enable learners to learn from and engage with international scholars and experts, thus forming an international community of learners dedicated to the understanding and meaningful internalization of EC in educational contexts. The final assignment of the course, preparing an EC educational intervention, focuses on student collaboration, teamwork, and comparative assessments by also engaging with different socio-cultural perspectives. This will allow students to develop practical toolkits and know-how for promoting and apply EEC in different social contexts.

Table 1. EEC Course structure.

Topic of Unit/Module	Teaching Methods	Example of Activity/Student Tasks
Course presentation and expectations Acquaintance	Face-to-face	Preparatory task: Intuitive perceptions of citizenship. What is your definition of civic/citizenship education? What does being a citizen mean for me today?
Framing the discourse: • Overview of the sustainability crisis (environmental–social); COVID-19 as a case study; VUCA world and wicked problems. • Sustainable development; viewing SD goals through the lenses of EC and sustainability. Citizenship for a VUCA world: EC/EEC from the perspective of key competences for lifelong learning (case: climate change)	Asynchronous	Readings: UN, 2015; UNESCO 2017; Capra and Luigi Luisi, 2014, OECD, 2018 (key competences) Create concept map of selected sustainability issue; map newspaper article of local/global sustainability issue.
Ecological footprint: Connecting lifestyles to sustainability issues	Asynchronous; collaboration with intl. students	Calculate personal/family/etc. EF using EF calculator and using a critical, comparative perspective of barriers and enablers for making change in relation to different cultural and social-institutional contexts.

Table 1. *Cont.*

Topic of Unit/Module	Teaching Methods	Example of Activity/Student Tasks
Environmental/Sustainability Citizenship: Theories of EC—how citizenship theoreticians frame sustainability in the context of citizenship (i.e., greening the concept of citizenship).	Asynchronous	Online interview with citizenship theoretician on EC in the context of citizenship theories Team task: Readings of citizenship theories and peer teaching Recorded lecture on EC—sustainability change agent framework (Sarid and Goldman, 2021), and EEC conceptualized within the ENEC.
The ENEC framework for EC; connecting EC and change agency as developmental constructs.	Asynchronous	
Situating EEC in contemporary educational theory: Competences for lifelong learning (OECD, Israeli Ministry of Education)	Asynchronous	Movie: OECD—Education 2030 (Compass clip) Israeli Ministry of Education (2020) framework for K-12 graduates' learning outcomes (knowledge, skills, and values) Team Task: Select pedagogy from a list as an example of a pedagogy conducive for promoting EC and prepare presentation for peer learning. Meet the experts: online recordings of lectures by international pedagogical experts together with MA students in international university
Selected progressive pedagogies associated with EEC (e.g., emancipatory learning, place-based, critical pedagogy, transformative learning, inquiry-based).	Collaboration with international lecturers	
Educational change agents toward EC: Workshops for developing educational interventions devoted to incorporating EEC in different educational contexts.	Face-to-face workshops (after the semester ends)	Final Product: Develop (in teams) work-plan for an educational intervention for incorporating EEC in specific cultural and educational context. International cooperation through either (a) mini-conference (long-distance) presenting interventions; (b) joint development in international teams via mobility.

Author Contributions: A.S. and D.G. contributed equally to all aspects of the conceptualization, writing—original draft preparation, writing—review and editing, and visualization. All authors have read and agreed to the published version of the manuscript.

Funding: This research received no external funding.

Institutional Review Board Statement: Not applicable.

Informed Consent Statement: Not applicable.

Data Availability Statement: Not applicable.

Acknowledgments: This study was partly inspired by COST Action ENEC—European Network for Environmental Citizenship (CA 16229), which is supported by COST (European Cooperation in Science and Technology).

Conflicts of Interest: The authors declare no conflict of interest.

References

1. OECD. Preparing our Youth for an Inclusive and Sustainable World–The OECD PISA Global Competence Framework. 2018. Available online: https://www.oecd.org/education/Global-competency-for-an-inclusive-world.pdf (accessed on 7 April 2022).
2. United Nations. Transforming Our World: The 2030 Agenda for Sustainable Development. 2015. Available online: https://sustainabledevelopment.un.org/content/documents/21252030%20Agenda%20for%20Sustainable%20Development%20web.pdf (accessed on 7 April 2022).
3. Wals, A.E.J. *Exploring Pathways to Sustainable Living: Emancipatory Environmental Education*; Area Studies (Regional Sustainable Development Review): Paris, France, 2009; pp. 209–229.
4. Cao, B. *Environmental Citizenship*; Routledge: London, UK; New York, NY, USA, 2015.
5. European Network for Environmental Citizenship—ENEC. 2018. Available online: http://enec-cost.eu/ (accessed on 7 April 2022).
6. Hadjichambis, A.C.; Paraskeva-Hadjichambi, D. Education for environmental citizenship: The pedagogical approach. In *Conceptualizing Environmental Citizenship for 21st Century Education*; Hadjichambis, A.C., Reis, P., Paraskeva-Hadjichambi, D., Cincera, J., Boeve-de Pauw, J., Gericke, N., Knippels, M.C., Eds.; Springer: Cham, Switzerland, 2020; pp. 237–261.
7. Sarid, A.; Goldman, D. A Value-Based Framework Connecting Environmental Citizenship and Change Agents for Sustainability—Implications for Education for Environmental Citizenship. *Sustainability* **2021**, 13, 4338. [CrossRef]

environmental sciences proceedings

MDPI

Proceeding Paper

Fostering Education of Environmental Citizenship through Food Living Labs †

Danielle Wilde [1,2,*] and Mary Karyda [1]

[1] Sociology, Environmental and Business Economics, University of Southern Denmark, 6705 Esbjerg, Denmark; mkaryda@sdu.dk
[2] Umeå Institute of Design, Umeå University, SE-901 87 Umeå, Sweden
* Correspondence: d@daniellewilde.com or danielle.wilde@umu.se

† Presented at the 2nd International Conference of International Researchers of the Education for Environmental Citizenship 2022, 10–11 March 2022. Available online: https://enec-cost.eu/ireec22/.

Abstract: The human food system is complex; has significant social and environmental impact; and raises questions around identity and culture. This mix of individual, collective, public, private and environmental concerns, positions Environmental Citizenship central to food system transformation. We discuss three 'FUSILLI' food living labs—a food waste NGO; a venue for creative experimentation of alternative food practices; and a forest-based library. These living labs use participatory research through design to place citizens at the forefront of change processes. We analyse them using the model of Education for Environmental Citizenship to consider how they foster EC and thereby sustainable food system transformation.

Keywords: EEC model; Food Living Labs; participatory research through design

Citation: Wilde, D.; Karyda, M. Fostering Education of Environmental Citizenship through Food Living Labs. *Environ. Sci. Proc.* **2022**, *14*, 22. https://doi.org/10.3390/environsciproc2022014022

Academic Editors: Andreas Ch. Hadjichambis, Pedro Reis, Marie-Christine Knippels, Demetra Paraskeva-Hadjichambi, Jan Cincera and Kateřina Jančaříková

Published: 21 June 2022

1. Introduction

Environmental citizenship (EC) is: "responsible pro-environmental behaviour of citizens who act and participate in society as agents of change..." [1]. We posit EC is fundamental to food system transformation. The human food system generates 20–40% of anthropogenic emissions [2]. The effects of climate change on food systems are extensive, complicated, geographically varied and seriously impacted by socioeconomic circumstances [3–5]. In such a critical and complex system, involving citizens in decision making as agents of change is crucial [6].

We discuss three food living labs developed in Kolding, Denmark, within the FUSILLI project [7]. FUSILLI has as its aim to transform twelve European city food systems, to be more sustainable through the implementation of innovative food living labs. The living labs we highlight include: a volunteer-run NGO that problematises food surplus distribution and waste; a food lab for experimentation and prototyping new food practices; and a Forest library that aims to entangle citizens with more-than-human nature.

To shape and evaluate the effectiveness of these efforts in developing EC and fostering food system transformation, we use the Education for Environmental Citizenship (EEC) model [8] to determine their impact. Our aim is to determine in what ways these food living labs foster EC, and whether there might be gaps in their development that warrant attention.

2. Materials and Methods

Food Living Labs are open innovation ecosystems that support experimentation, collaboration and learning around food system transformation [7]. The term "living lab" [9] indicates a methodology where people formulate, prototype, and substantiate complex solutions in real-life environments. Food Living Labs are public and communal, whereas Citizenship is often experienced as individual. According to the EEC model [8], the actions that a citizen can take have individual and collective dimensions and may be applied in private and public spheres on local, national and global scales. The model highlights personal capacities related to EC: knowledge, values, attitudes, skills, competencies and behaviours, and introduces the notion of being an agent of change, who participates actively in decision-making processes. We use the EEC Model to determine the relationships between the presented Food Living Labs and the need to foster EC to achieve sustainable food system transformation.

3. Food Living Labs as Incubators for Environmental Citizenship

We describe the living labs, then use the EEC model to formulate new understandings about how they enable citizens to not only participate or be responsible but to develop their capacities as agents of change (moving through different forms of citizenship [10] towards a full expression of EC). The living labs are in different stages of development. They show how FUSILLI gathers stakeholders, develops new forms of governance, raises awareness of nature-relatedness and finds ways of transforming impacts across the food value chain. Each serves as a platform for stakeholders to engage with problems and activate change processes using experimentation and dialogue with food as the primary enabler, approaching food as subject, research object, cultural practice and vibrant matter (following [11]).

Food Reformers [12] (est. June 2019) is a volunteer-run NGO that brings focus to surplus food, to raise awareness around food waste and distribution and press for food system change. The NGO hosts regular 'Dumpster Dinners' in a local restaurant, using donated and rescued surplus food (from local supermarkets). In June 2021, they partnered with R10, a local citizen involvement initiative [13], to build a Community Fridge to expand their impact. The fridge is centrally located and open 24/7 [14]. It contains rescued food, in good condition, with an impending sell-by date. As an organisation, Food Reformers is nationally certified to redistribute the collected food. They are active in the community and promoted by the municipality and other organisations, which inspires confidence. Their work has been introduced nationally by themselves and like-minded activists such as Matt Homewood [15], and at the European level and globally, through FUSILLI.

Food Lab [16] (est. November 2021) sits in the centre of Kolding, at R10, where the Community Fridge is found. Food Lab is a venue for experiments, events and workshops focused on participatory governance and citizen innovation actions across the food value chain. These efforts use food as the medium for experimentation to prototype new ways of thinking about the food system transformation. At the Food Lab soft launch, we served surplus food collected by Food Reformers, in the form of relevant data— e.g., a guacamole with strings threading out the relative distance travelled by each ingredient—to make tangible the challenges we must grapple with in the food system. Guests included politicians, municipal representatives, educators, entrepreneurs, farmers, representatives of the water and energy companies, a rewilding consultancy, a poet and more. They enjoyed the food while using it as a ticket-to-talk [17] and a prompt for a facilitated discussion around four questions: What does a sustainable food system look like to you? What do you need to be sustainable in your food-related practices? What do wildlife need to flourish? Where can we start? At the subsequent public launch, we held workshops on i) Food Waste and Packaging, and ii) Aquaponics in small spaces and mental health. Working with citizens, we established a wildflower garden and mini forest, which serve as an exemplar pollinator restaurant and hotel, for a forthcoming citizen science project.

Forest Library (in the making). In September 2021, an ad-hoc Steering Committee (SC) was tasked with establishing a forest-based Library of Foraged Foods and Practices to support citizens to gain knowledge about what grows locally; what can be foraged from forest, field and fjord; how to connect to, respect and engage with nature; sharing knowledge in more-than-human exchange. The SC includes representatives from research and education, small businesses, the Art and Design Museum, chefs, librarians and more. To date, two meetings have been held in the proposed forest location: a walk-and-talk and (10-days later) a dinner. The walk-and-talk introduced the SC to the forest, each other and our partially formed ideas and provided space to negotiate a collective vision for the library. The dinner included a backcasting session using found objects for envisioning and planning [18] and a meal of elements foraged from the forest, brought together by a local chef. These activities resulted in a co-created mapping of concerned stakeholders; revealed a map of interconnections between divergent agendas; helped to consolidate the vision for the library; and articulate a burgeoning plan for its implementation.

4. Findings

Figures 1 and 2 maps the living labs to the EEC model, differentiating between the actions of the living labs, participating citizens and organisational stakeholders. For analysis, the Forest Library is separated into formation efforts and vision. In each case, the living labs curate collective efforts in the public sphere, typically hyper-local actions that draw on participatory research through design methodologies [18] to invite enriched, embodied participation. Careful attention is paid to expanding participants' capacities for EC, attending to the characteristics at the centre of the EEC model, which are understood to scaffold the pathway towards being agents of change. The mapping affords nuanced consideration of each living lab.

FoodReFormers' efforts (Figure 2, top-left) take place at a local scale, supported by a solid infrastructure of volunteers and other stakeholders (e.g., the Municipality). Their impact extends beyond the local by telling their story on international stages. The strong establishment of local output relates to how long that living lab has been operating, combined with consistently solid infrastructure and branding. In contrast, Food Lab (Figure 2, top-right), was established less than a month ago, yet has a strong research foundation. Despite being in its early stages, its national and global output is already well established. This difference demonstrates two very different pathways towards EC from an infrastructure perspective.

With Forest Library, we see established activities (Figure 2, bottom-left) tightly held to collective actions in the public sphere. The vision—the propositional mapping (Figure 2, bottom-right)—has expansive reach across spheres and scales. Similar to the Food Lab, the Library is supported by a strong research foundation. As with Food Reformers, it represents a singular vision, driven by local stakeholders. The missing link, for the moment, is strong involvement of those stakeholders to achieve the level of societal integration demonstrated by Food ReFormers, which has a constant flow of citizens interested in knowing more about their work and output. The qualities of engagement they provide through their interventions activate people in wanting to join their efforts. Those techniques of engagement—future workshops and social interventions that leverage commensality—are at the foundation of all three living labs, and we hope will result in similar impact, whether their vision is singular, or—as in the case of Food Lab—embodies a multi-faceted commitment to the full food system value-chain.

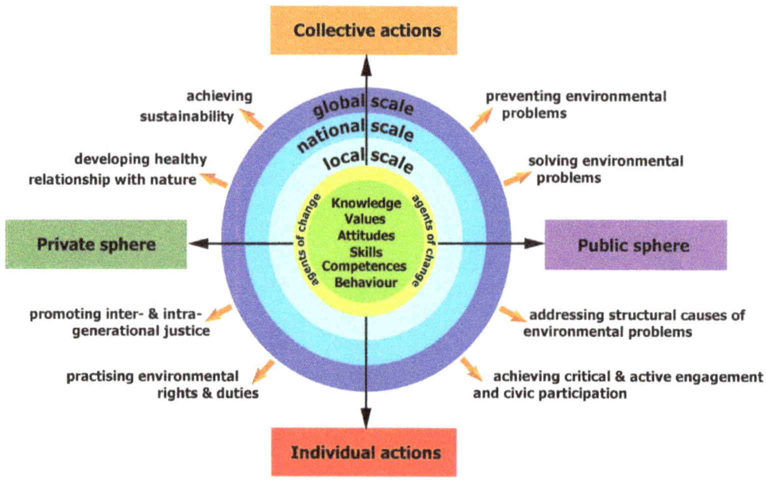

Figure 1. The Education for Environmental Citizenship (EEC) Model [8:240].

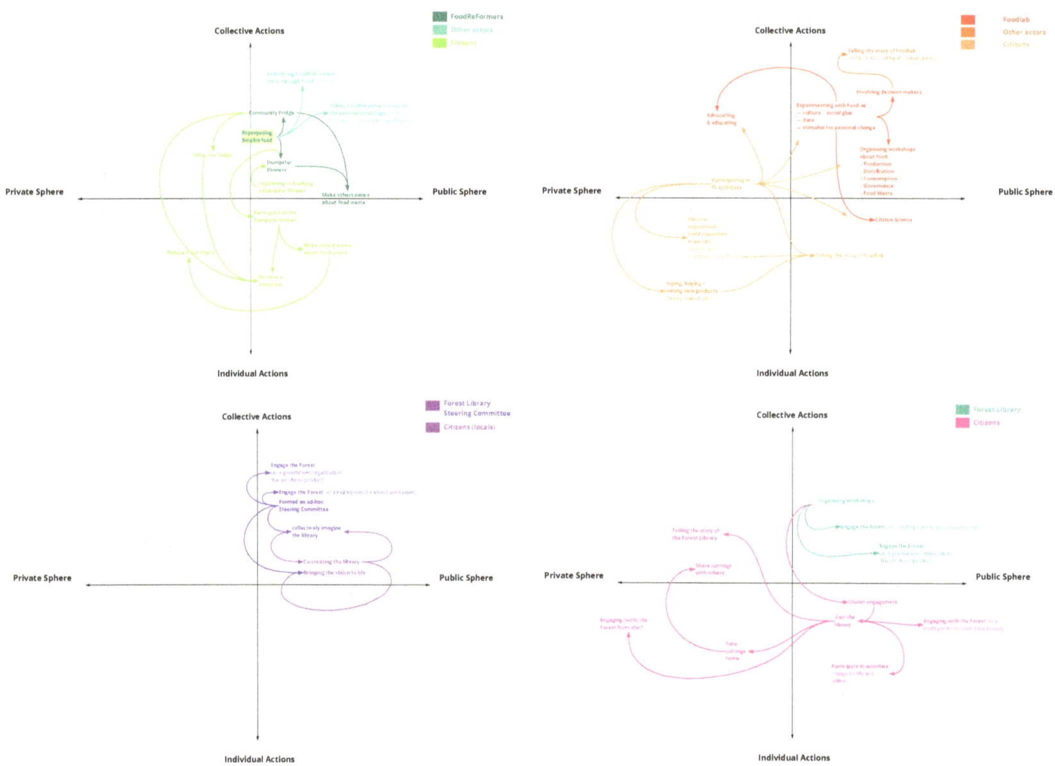

Figure 2. Three FUSILLI Food Living Labs mapped to the EEC model: top-left: Food Reformers; top-right: Food Lab; bottom-left: Forest Library Steering Committee efforts; bottom-right: Forest Library vision.

5. Conclusions

The three FUSILLI Living Labs we described here draw on participatory research through design to involve diverse citizens in food system transformation. They use food as the locus for radically open innovation that both requires and fosters EC. The result is a form of non-formal education that has a certain aesthetic quality and playful elements which invite for constant participation, and enriched engagement with the burning topic of food system transformation. When analysed according to the EEC model, this approach seems to scaffold the skills that citizens of diverse ages and interests require to participate in society as agents of change, individually and collectively, in public and private spheres, with local, national, and international impact.

Author Contributions: Conceptualization, D.W.; methodology, D.W.; formal analysis, D.W. and M.K.; investigation, D.W. and M.K.; writing—original draft preparation, D.W.; writing—review and editing, D.W. and M.K.; visualization, D.W.; supervision, D.W.; project administration, D.W.; funding acquisition, D.W. All authors have read and agreed to the published version of the manuscript.

Funding: This project has received funding from the European Union's Horizon 2020 research and innovation programme under grant agreement No. 101000717.

Institutional Review Board Statement: The study was performed under ethical guidelines from the Faculty of Humanities of The University of Southern Denmark; explicit ethical review for this study was not mandatory.

Informed Consent Statement: Informed consent was obtained from all subjects involved in the study.

Data Availability Statement: Data sharing not applicable.

Conflicts of Interest: The authors declare no conflict of interest.

References

1. European Network for Environmental Citizenship. Available online: https://enec-cost.eu (accessed on 18 December 2021).
2. Rosenzweig, C.; Mbow, C.; Barioni, L.G.; Benton, T.G.; Herrero, M.; Krishnapillai, M.; Liwenga, E.T.; Pradhan, P.; Rivera-Ferre, M.G.; Sapkota, T.; et al. Climate change responses benefit from a global food system approach. *Nature Food* **2020**, *1*, 94–97. [CrossRef]
3. Masson-Delmotte, V.; Zhai, P.; Pirani, A.; Connors, S.L.; Péan, C.; Berger, S.; Caud, N.; Chen, Y.; Goldfarb, L.; Gomis, M.I.; et al. (Eds.) *IPCC, 2021: Climate Change 2021: The Physical Science Basis. Contribution of Working Group I to the Sixth Assessment Report of the Intergovernmental Panel on Climate Change*; Cambridge University Press: Cambridge, UK, 2021; Available online: https://www.ipcc.ch/report/ar6/wg1/ (accessed on 18 December 2021).
4. Vermeulen, S.J.; Campbell, B.M.; Ingram, J.S. Climate change and food systems. *Annu. Rev. Environ. Resour.* **2012**, *37*, 195–222. [CrossRef]
5. Willett, W.; Rockström, J.; Loken, B.; Springmann, M.; Lang, T.; Vermeulen, S.; Garnett, T.; Tilman, D.; DeClerck, F.; Wood, A.; et al. Food in the Anthropocene: The EAT- Lancet Commission on healthy diets from sustainable food systems. *Lancet* **2019**, *393*, 10170. [CrossRef]
6. Wilde, D.; Hupe, A.L.; Trahan, S.; Guinita Abel, C.; Kjærsgaard Longueval, S.; McLaughlin, C. Rethinking Food: Co-Creating Citizen Science for Sustainability Transitions. In *Nordes 2021: Matters of Scale, Proceedings of the Nordes International Conference. Kolding, Denmark, 15–18 August 2021*; Brandt, E., Markussen, T., Berglund, E., Julier, G., Linde, P., Eds.; Designskolen Kolding: Kolding, Denmark, 2021; pp. 228–237.
7. Braathen, E.; Wilde, D.; Borgen, S.O.; Eika, A.; Karyda, M.; Søvik, A. 2021. Living Labs for Urban Food System Transformation an Inventory Report. Fostering the Urban Food System Transformation through Innovative Living Labs Implementation. Research & Innovation Action. CE-FNR-07-2020: FOOD 2030 Empowering Cities as Agents of Food System Transformation. Available online: www.fusilli-project.eu (accessed on 18 December 2021).
8. Hadjichambis, A.; Paraskeva-Hadjichambi, D. Education for Environmental Citizenship: The Pedagogical Approach. In *Conceptualizing Environmental Citizenship for 21st Century Education*; Hadjichambis, A.C., Reis, P., Paraskeva-Hadjichambi, D., Činčera, J., Boeve-de Pauw, J., Gericke, N., Knippels, M.C., Eds.; Springer Nature Environment Discourses in Science Education: Berlin, Germany, 2020; Volume 4, pp. 237–261. [CrossRef]
9. Eriksson, M.; Niitamo, V.P.; Kulkki, S. *State-of-the-Art in Utilizing Living Labs Approach to User-Centric ICT Innovation a European Approach.* Center for Distance-Spanning Technology, Lulea University of Technology, Lulea, Sweden, 2005. Available online: http://84.88.32.6/openlivinglabs/documents/SOA_LivingLabs.pdf (accessed on 19 December 2021).
10. Westheimer, J.; Kahne, J. What kind of citizen? The politics of educating for democracy. *Am. Educ. Res. J.* **2004**, *41*, 237–269. [CrossRef]

11. Bennett, J. *Vibrant Matter: A Political Ecology of Things*; Duke University Press: Durham, NC, USA, 2010. [CrossRef]
12. Food ReFormers. Available online: https://www.facebook.com/foodreformer (accessed on 19 December 2021).
13. Riberdyb10. Available online: https://www.facebook.com/Riberdyb10 (accessed on 19 December 2021).
14. Fællesskabet Kolding Free Fridge Kolding. Available online: https://www.facebook.com/FaellesskabetKolding (accessed on 19 December 2021).
15. Matt Homewood. Available online: https://www.matthomewood.com (accessed on 19 December 2021).
16. Karyda, M.; Wilde, D.; Gislev, K.J.; Rsgaard, M. Narrative Physicalization: Supporting Interactive Engagement with Personal Data. *IEEE Comput. Graph Appl.* **2020**, *41*, 74–86. [CrossRef] [PubMed]
17. Food Lab. Available online: https://koldingfood2030.dk/index.php/food-lab-2/ (accessed on 19 December 2021).
18. Wilde, D. Design research education and global concerns. *She Ji* **2020**, *6*, 170–212. [CrossRef]

MDPI
St. Alban-Anlage 66
4052 Basel
Switzerland
Tel. +41 61 683 77 34
Fax +41 61 302 89 18
www.mdpi.com

MDPI Books Editorial Office
E-mail: books@mdpi.com
www.mdpi.com/books

www.ingramcontent.com/pod-product-compliance
Ingram Content Group UK Ltd.
Pitfield, Milton Keynes, MK11 3LW, UK
UKHW060125010826
14089UKWH00006B/152

9 783036 540719